Applied Fourier Analysis

Tim Olson

Applied Fourier Analysis

From Signal Processing to Medical Imaging

 Birkhäuser

Tim Olson
Department of Mathematics
University of Florida
Gainesville, FL
USA

ISBN 978-1-4939-8471-8   ISBN 978-1-4939-7393-4   (eBook)
https://doi.org/10.1007/978-1-4939-7393-4

Mathematics Subject Classification (2010): 42A38, 42B10, 65T50, 92C55, 94A08, 94A12, 94A20

Printed on acid-free paper

This book is published under the trade name Birkhäuser, www.birkhäuser-science.com
The registered company is Springer Science+Business Media, LLC
The registered company address is: 233 Spring Street, New York, NY 10013, U.S.A.

Preface

Purpose

This book is the outgrowth of many years of trying to quickly teach engineers and mathematics majors Fourier Analysis at University of Florida. The major constraints were the following:

1. The students will not necessarily have a prior course in modern analysis; therefore, Lebesgue measure is suppressed. Linear Algebra and calculus is all that is assumed.
2. The students generally do not have two or three semesters to master the subject. They have one semester to get an overview of the subject. While those are the constraints in Gainesville, I believe that this book will support a two-semester course taught slower with complete detail. In ten years, I have never covered everything in this book.
3. Most of the students will not be interested in an analysis class which does not cover some applications. Understanding some applications needs to be a goal of the class.
4. The course should have mathematical rigor, but this rigor should not be emphasized at the cost of applications.
5. The students should be exposed to the concepts of isometry, sampling, interpolation, and aliasing.
6. When I wrote the first test for this class, many of the questions had been on previous Numerical Analysis Ph.D. qualifying exam. Thus, I had to find a way to let undergraduate students learn, without expecting heavy proof ability as would be expected at a much later date.
7. Some exercises which involve the calculation of Fourier Series or integrals are necessary and helpful. Calculation of complicated integrals should not be the only emphasis of the course.

8. Projects seem to be the best way of having them "get their feet wet". Most notably numerical projects are using MATLAB, although other packages could suffice. I let them program in any language they wish, but refuse to answer questions about anything other than MATLAB. I have included some of the projects I used for one semester. Please feel free to alter them.

While there were many mathematics students in these courses, many of them are EE, CS, ME, Physics, etc. Thus, the purpose is to give a mathematical introduction to Fourier Analysis, while emphasizing the applications. The goal is not to give a final course on Fourier Analysis with complete mathematical rigor. The goal is not to make the course a final course on any of the applied subject topics. The goal is to set the plate for further investigation by the students into the topics which they find most interesting or which their individual study plans demand.

Suggestions to Professors

Covering the Material

One obvious observation is that there is a lot of material in the book for one semester. I may have come close to covering all of it in one semester, but that is never my goal. I would love to be at a place where the students do not need to graduate, and we could take two semesters to cover the material. In a thirteen-week semester, I generally spend 7–8 weeks on Chapters 2-4. I try to emphasize to the students my belief that if they know this material, they can read the rest easily. I do emphasize Chapter 5, since sampling is at the forefront of most of modern signal processing, communications, etc.

After Chapters. 2-4, things become far easier for the students, and generally, it is fun time to play with what they learned. I usually cover the later chapters in a week or so. I pick and choose them according to what the students are interested in. Normally, I move to Image Processing after Chapter 5, because nearly all students want to be able to play with images. The Medical Imaging chapter is of personal interest to me, and I find it a lot of fun. I believe that students do also. I generally end up with some cleanup time, i.e., just covering the highlights of chapters which we did not cover, sometimes one lecture per chapter.

In summary, I do not feel it is necessary to cover more than one or two chapters after the base knowledge in Chapters 2-5. If I have time, I keep going.

Projects

You will notice that the chapter projects are merely collections of exercises. There is no magic in how I have arranged them. They seem to work, but they are continually evolving. You may want to change small things because code from prior semesters will always hang around and is hard for students to resist. For instance, merely changing expansions on $[-\pi, \pi]$ to $[-2, 2]$ will insure that the students have to put some of their own thought into the projects.

It is completely viable to build custom projects from individual exercises. I tend to stay with ones that have traditionally worked. Whether other projects are designed is certainly up to the individual instructor.

Testing

Part of the reason I wrote this book was that testing on this broad material was difficult, and probably above the level of an undergraduate class. On the other hand, I do give one test per semester, generally over the material in Chapter 4. The purpose of that test is to make sure that the students know some of Fourier Analysis off the top of their head. Required material includes the basics of dilation, translation, differentiation, convolution, etc. I refer to these as the "easy" problems. I challenge them to know some of the "hard" problems, such as the uncertainty principle, the proof of the Inverse Fourier Transform, and the Shannon Sampling Theorem. I do not require all of them.

There is no magic here, just what I found necessary to feel that I had transferred the knowledge. Please experiment with a mix of testing vs. projects. If you find something that works, please let me know.

Acknowledgements and History

Undergraduate and Graduate Years

This book is somewhat of an oral history of my training in mathematics. I was an undergraduate student at University of Montana in Missoula. It was only through the efforts of some dedicated Professors, most notably George McRae and William Derrick, that I survived my time as an undergraduate with some academic integrity in place. They convinced me that I could succeed at Mathematics if only I decided to study.

I attended graduate school at Auburn University and began to look at Mathematics as a career and something that I loved, rather than a distraction from my other pursuits. The entire faculty at Auburn was very engaged in educating the graduate students, and I have to commend all of them. My thesis advisor Dr. Richard Zalik treated me as one of his family and guided me through my experience there. My experience at Auburn included a wide number of the Professors, and I would like to thank all of them. I was also blessed with a great number of excellent colleagues at Auburn, and I have to mention Saady, Bobby, and Jeffy, who received their Ph.Ds with me.

While at Auburn, I studied two rather different disciplines. I wrote my Master's Thesis on Tomography [20] and learned a great deal about dealing with real data and applied numerical analysis. I then wrote a fairly pure mathematical thesis on wavelet-type basis for Fourier Analysis [21]. These two different approaches, one very applied and one fairly pure, are hopefully illustrated in this book.

The Dartmouth Years

I was privileged to leave Auburn to work with Dennis Healy et al at Dartmouth on a postdoctoral fellowship. Dartmouth was a great place to be at the time. Dennis and I worked on a number of applications, medical imaging, radar processing, statistical pattern recognition, as well as others.

I also had the blessing of working with a great group of others, including Dan Rockmore and John Weaver. We eventually named our group the "Wavelet Warriors." While we rarely did anything directly with wavelets, we believed that Fourier Analysis, appropriately used through wavelets or otherwise, could solve problems. We solved some, and we left many unsolved, but it was a great time and place to work. Most of the Wavelet Warriors are shown in Figure 1, including the wavelet dogs, Charlie Olson and Digger Rockmore. While at Dartmouth, we were a family that generally played and worked together. I will always remember it as a very special time.

University of Florida

I spent a year at Johns Hopkins, studying under Carey Priebe. We enjoyed the time and he enlightened me to the wonders of spatial statistics. I have been at University of Florida since 1997. My colleagues have all been great. I must acknowledge the great luck I had to first study at Dartmouth with the Wavelet Warriors, and then to end up at Florida. John Klauder has done his best to keep my knowledge of Fourier Analysis alive and keep me out of trouble. Both are full-time jobs. Sergei Shabanov has tolerated my ignorance of physics and answered any questions I had.

Figure 1: Wavelet Warriors are depicted in this photograph from June of 1996. In the top row are Peter Kostelec, Doug Warner, Sumit Chawla, Geoff Davis, and Dennis Healy. At the bottom are Dan Rockmore, Digger Rockmore, Charlie Olson, and Tim Olson. The Wavelet Warrior Viking regalia is circa 1995 and brought much strength to the group.

Dedication:

This book is dedicated to Professor Dennis Healy and the other Wavelet Warriors, at Dartmouth and beyond. Professor Healy unfortunately departed this world in September of 2009, but he will be with all of us forever.

Contents

Chapter 1
Introduction: From Linear Algebra to Linear Analysis

The subject of Fourier Analysis is at the center of many modern sciences. Physics and engineering depend so heavily on the subject, it is hard to imagine even mentioning them without thinking immediately about frequency-dependent phenomena: light, sound, orbits, and vibrations to name a few.

Many other scientific fields also are highly dependent upon Fourier Analysis for understanding even the most basic of phenomena, such as population dynamics in biology. Fourier Analysis is also essential in understanding what are some of the most common occurrences in everyday life. The various periodic phenomena of the sun, the moon, the tides, and the seasons are yet to be completely understood. We know the sun and moon will rise with certainty, and their secondary seasonal effect on wind and tide-related phenomena such as el Nino are still somewhat of a mystery. These periodic effects are drastic and life altering, and are naturally studied with Fourier Analysis.

Before we embark on trying to solve all of the world's scientific mysteries with Fourier Analysis, let us stop asking a simple question. *"Why Fourier Analysis? Should I spend a great deal of time trying to understand this subject? Why should I worry about Fourier Analysis?"*

The simple answer to this question is outlined above. Many of the important phenomena in our world are related to periodic phenomena, such as the rising of the sun and the moon. The idea of frequency is so basic that we sometimes have a hard time deciding which version to work from. The fact that sound is a periodic, frequency-based phenomena adds to this. The discovery that light is frequency based is a relatively new idea, perhaps only well understood for the last 300–400 years.

Mathematically, I like to say that in a vast majority of cases, sines and cosines or close relatives of them are the eigenfunctions, or building blocks of the universe. If you understand them well, and how they interact with different systems, you will be able to describe and understand a great deal of modern science. The goal of this book is to have the reader understand

© Springer Science+Business Media, LLC 2017
T. Olson, *Applied Fourier Analysis*,
https://doi.org/10.1007/978-1-4939-7393-4_1

that *"Yes, an understanding of Fourier Analysis is a very worthwhile goal."* I believe that most modern scientists would agree with this statement. Exactly how much time and how deeply you explore the subject of Fourier Analysis may depend upon your specialty and motivations.

1.1 Three variations of Fourier Analysis

We will concentrate on three important variations of Fourier Analysis in this book: (1) Fourier Series, (2) The Discrete Fourier Transform, and (3) The Continuous Fourier Transform. The interplay between these three structures is crucial to understanding the whole of Fourier Analysis. They are very similar with crucial differences.

Fourier Series involves representing a continuous function on a finite interval $[a, b]$ with an infinite series which is similar to a Taylor series. This is the transform which historically came first and from which the other two structures grew. The Discrete Fourier Transform concentrates on a function on a finite set of points, which we will refer to as $0, 1, \ldots N-1$. The Discrete Fourier Transform is the version which is most commonly implemented in computers, in cell phones, etc. The Continuous Fourier Transform is an extension of Fourier Series and is necessary to understand the deep connections between physical reality and the digital representations or approximations which we can generate with Fourier Series or the Discrete Fourier Transform (DFT).

1.1.1 Fourier Series

A simple form of Fourier Analysis is the basic Fourier Series. Simply stated a function on the interval $[-\pi, \pi]$ can be represented in a series expansion

$$f(t) = \frac{a_0}{2} + \sum_{k=0}^{\infty} a_k \cos(kt) + b_k \sin(kt),$$

where

$$a_k = \frac{1}{\pi} \int_{-\pi}^{\pi} f(t) \cos(kt) dt, \quad \text{and} \quad b_k = \frac{1}{\pi} \int_{-\pi}^{\pi} f(t) \sin(kt) dt.$$

Thus, we can represent a function in terms of individual sine and cosine terms. We have now begun the most basic form of Fourier Analysis, which is to separate a function into the various frequencies which represent it. If you think of sound, these are the most basic things we hear. If this is voice, the high frequencies will generally dominate a child's or female's voice. Lower frequencies generally are associated with an adult male's voice.

There is a secondary advantage of the above series representation. The cosine terms are even, and the sine terms are odd. Therefore, we have also decomposed our function into even and odd parts, in addition to the individual frequencies. This will prove very valuable for a number of applications, such as compression and partial differential equations.

The complexities of these frequency combinations become obvious when we realize that we can generally recognize our relatives and good friends from the composition of their voice or the various frequencies in their voice.

In music, these frequencies will approximately correspond with the notes and instruments which are present. In an orchestra, violins are higher frequencies and violas and basses are lower frequencies.

One limitation of the Fourier Series can be seen in the formulas above. You must calculate the integrals, which is usually time consuming and oftentimes not practical for real-world signals. A second limitation is that a Fourier Series is only valid on a finite interval. This is a good and a bad thing. Frequencies are oftentimes only present for a short period of time. Therefore, assuming that they continue forever is not representative of our normal experience with voice or music. Trying to find a representation which locates frequencies and the time periods in which they exist is surprisingly challenging and very rewarding.

1.1.2   The Discrete Fourier Transform

We noted above that Fourier Series requires the calculation of integrals, (an infinite number of them), and that is not always feasible. The Discrete Fourier Transform (DFT) addresses this problem by creating a simple and wonderfully structured matrix which can be used on basic finite signals. Suppose that you have a function $f(t)$ on some interval, and you want to represent it with N samples at discrete time points t_n, where $n = 0 \ldots N - 1$. We will refer to this vector as $\vec{f}_n = \{f(t_n)\}_{n=0}^{N-1}$.

The Discrete Fourier Transform actually samples the cosine and sine terms, so the vector representing $\cos(t)$ would be $\cos(2\pi n/N)$, where $n = 0 \ldots N-1$. Note that this vector $\cos(2\pi n/N)$ goes through one cycle of the cosine just as $\cos(t)$ would go through one cycle on $[0, 2\pi]$. We then use these sine and cosine vectors to construct the Fourier matrix F_N where N represents the number of points in our sampled function. By doing this carefully, we can then write the vector which we sampled from our function $f(t)$, $\vec{f}_n$ as a multiple of these cosine and sine vectors, or in the form

$$\vec{f} = F_N \vec{c}.$$

where c are the coefficients of our sampled sines and cosines. We will also set this up, so F has a simple inverse or so that we can easily find $\vec{c}$ with the formula

$$\vec{c} = F_N^{-1} \vec{f_n}.$$

The obvious advantage of this approach is that we do not have to calculate integrals. We can simply use Linear Algebra. A bonus is that we can decompose the matrix F_N in a special way which will make the Linear Algebra extremely fast. This is a crucial enabling factor in the digital world in which we live.

There are several questions and problems which we must address to utilize this approach: (1) How often do we sample, or how close do the points t_n have to be so that we can accurately study the behavior of $f(t)$ from a finite number of samples $f(t_n)$? (2) Do we utilize the above matrix on short portions of the function, or long portions? (3) Will the way we use this represent the real frequency content of the original function? (4) Can we calculate these expansions quick enough to be useful in applications such as real-time communication systems? These are all questions which have answers, but they are not all solvable in a simple way.

1.1.3 The Continuous Fourier Transform

The story of Fourier Analysis would not be complete without the Continuous Fourier Transform or simply the Fourier Transform. We sometimes compute the actual Fourier Transform on simply defined functions. More generally, we utilize the lessons and theory which are learned by studying the Fourier Transform to compute Fourier Series and the DFT properly. We can answer many of the problems with the Discrete Fourier Transform and Fourier Series with the Fourier Transform.

The coefficients involved in Fourier Series are actually subsamples of the Fourier Transform. We define the Fourier Transform as

$$\hat{f}(s) = \frac{1}{\sqrt{2\pi}} \int f(t) e^{ist} dt = \frac{1}{\sqrt{2\pi}} \int f(t) \left(\cos(st) + i \sin(st) \right) dt.$$

Note that the first part of the integral on the right has the formula for the a_k coefficients of the cosine coefficients from the Fourier Series as a subset. The second portion of the integral on the right has the sine or b_k coefficients from Fourier Series as a subset.

These discrete subsets or samples of the Fourier Transform generate sampling theory which we will study extensively. Understanding exactly how many discrete samples are enough to correctly represent a function is crucial.

In addition, there are a number of other beautiful phenomena which are most easily understood and expressed via the Fourier Transform and its major theorems, such as the uncertainty principle.

1.2 Motivations

There are a number of reasons why one would want to use Fourier Analysis to study, solve, or manipulate the results of a problem or process. Oftentimes those of us who use Fourier Analysis on a daily basis do not really think about what type of motivation we have for using the tools at our disposal. Let us try to give a brief overview of a few different types of approaches to utilizing the Fourier Transform.

1.2.1 Exploration and Understanding

The most basic goal of science is to give a simple explanation of a physical, biological, or general phenomenon. There are a wide variety of physical phenomena which can be explained quickly with sines and cosines. The motivation is to solve a problem, or understand the problems which come from a basic system.

A simple physical model, which is oftentimes applicable to physics, engineering, biology, or chemistry, is a set of ordinary differential equations which models the underlying process. Oftentimes, we utilize an equation which has a second derivative as its highest derivative, and the general model of this type is

$$ay''(t) + by'(t) + cy(t) = f(t).$$

We will assume for now that $f(t) = 0$, which means that no outside forces are acting on the model.

The standard way to solve this equation is to "guess" that solutions may be of the form $y = e^{\lambda t}$. This implies that $y'(t) = \lambda e^{\lambda t}$ and $y''(t) = \lambda^2 e^{\lambda t}$. Substituting this possible solution and derivatives into the equation, we get

$$a\lambda^2 e^{\lambda t} + b\lambda e^{\lambda t} + ce^{\lambda t} = 0.$$

Since $e^{\lambda t} \neq 0$ for any t, we may factor this out to find that our solution depends upon

$$a\lambda^2 + b\lambda + c = 0.$$

Thus, we have changed a differential equation into a simple polynomial equation. More importantly, we can easily solve this equation with the quadratic formula to get

$$\lambda = \frac{-b \pm \sqrt{b^2 - 4ac}}{2a}.$$

At this point, it is still not clear why Fourier Analysis has anything to do with this example. The key factor in the above quadratic formula is the discriminant, or $b^2 - 4ac$. If $b^2 - 4ac > 0$, then the above solutions are both real, and the corresponding solutions to the differential equation will exhibit either exponential growth or decay. Exponential growth is typical of a bacteria in a petri dish with plenty of resources and no competing organisms. Exponential decay is common in a situation where a signal is being absorbed as it transmits through a medium.

Fourier Analysis comes into play when $b^2 - 4ac < 0$. In this case, the λ's will be conjugate pairs of one another, which we will denote by $\lambda_\pm = \alpha \pm i\beta$, where $i^2 = -1$. This leads to the complex exponential solutions

$$e^{\lambda_{pm} t} = e^{(\alpha \pm i\beta)t} = e^{\alpha t} \left(\cos(\beta t) + i\sin(\beta t) \right). \tag{1.1}$$

Generally, it is assumed that $\alpha \leq 0$. If $\alpha = 0$, then we have simple sinusoidal solutions. This is generally not realistic since the solutions would continue to oscillate forever. The more realistic solutions occur when $\alpha < 0$, which means that the solution to the differential equation decays exponentially to zero, with continual oscillation along the way. If $\alpha > 0$ on the other hand, the solution would explode, which is usually not consistent with a physical system.

Figure 1.1:   Above you see the Tacoma Narrows Bridge in Washington state, also known as Galloping Gertie, beginning to self-destruct due to resonant frequencies induced by the wind in 1940. You can find video of this in many places.

Fourier Analysis is obviously concerned with the vibrational or oscillating component of 1.1, or the terms $\cos(\beta t)$ and $\sin(\beta t)$. The frequency β is oftentimes referred to as the resonant frequency of the system. From the viewpoint of a mechanical engineer, it is very important to know what the resonant frequency of a system is. If the resonant frequency of a mechanical system, such as a bridge, can be induced by nature or an external force, then the system will absorb that energy and eventually may catastrophically destroy itself.

This was the case of "Galloping Gertie," which is the retroactive name for the Tacoma Narrows Bridge in Washington which is illustrated in Figure 1.1. Online video of this disaster can be found with a simple search and is quite interesting. The bridge was destroyed in 1940 because crosswinds excited the bridges' resonant frequencies caused it to vibrate, or gallop, until it collapsed. Obviously, mechanical engineers have been very concerned about dominant frequencies ever since. Understanding the relationships between frequencies and systems, mechanical, physical, biological or otherwise, is one basic goal of Applied Fourier Analysis.

1.2.2   Manipulation

We always want to understand the basics of any system which we are studying. Oftentimes, especially in engineering, we want to use this knowledge to manipulate the data gathered from a system. There are a number of reasons why one might like to manipulate the output or behavior of a system, such as: (1) Removing background noise or static from a voice signal so that individual conversations can be more clearly heard, (2) Emphasizing major long-term trends in a system, which are sometimes clouded by a number of minor and less important short-term phenomena, and (3) Being able to clearly enhance radar signals and their displays so that air traffic controllers can safely direct operations in the air at a major airport.

Let us pick one simple example of how one might enhance a simple radio signal.

1.2.3   Compression

Compressing information becomes more and more important in the current digital age. Cell phone traffic, as well as real-time video and sharing of a tremendous number numerous pictures, has become extremely common. In addition, video stores such as Blockbuster, which were fixtures of society 10 years ago, have gone by the wayside. We can now download full movies instead of having to go to the video store. Digital telephones now take extremely high-resolution pictures which are then compressed for transmission, nearly instantaneously.

At first glance, this application does not seem to have anything to do with Fourier Analysis. We will learn that Fourier Analysis is very essential to this basic task. We will show that simple changes to the way things are represented and stored can save vast amounts of storage space. Compression is not only needed for storage, however. Compression is perhaps far more valuable for communications, where one needs to transmit video or audio over a communications channel. Virtually all types of communication have costs associated with them which are directly tied to the amount of data transmitted. If you are transmitting wirelessly, you have to pay the Federal Communications Commission for your airspace. The leases for this nationally are incredibly expensive. In fiber, or ordinary copper cable, additional bandwidth requires switching stations, etc. Thus, the compression information without losing quality, transmission, and restoration of the information is very valuable to many people.

There many reasons why an average home now can receive 400–500 channels of TV, some of which is HDTV, on the same cable that supported perhaps 40 two decades ago. One primary reason is the efficiency of compression algorithms, many of which come from Fourier Analysis.

Problems and Exercises:

1. Use the Taylor series expansion of e^x to show that

$$e^{i\beta t} = \cos(\beta t) + i\sin(\beta t).$$

2. Differentiate $e^{i\beta t}$ using the rules for differentiating exponentials, and then differentiate $\cos(\beta t) + i\sin(\beta t)$ using the rules for differentiating sines and cosines. Do you get the same thing?

3. Differentiate $e^{\alpha t + i\beta t}$ using the rules for exponentials. Use the product rule to differentiate $e^{\alpha t}(\cos(\beta t) + i\sin(\beta t))$. Do they correspond? Is this sufficient to conclude they are the same?

1.3 Linear Algebra, Linear Analysis, and Fourier Analysis

Fourier Analysis might seem like a very exotic subject, which you have never seen before. The reality is that it is just an extension of Linear Algebra which most students should be familiar with. "What does Linear Algebra have to with Fourier Analysis?" The answer is simple. Fourier Analysis is the study of sines and cosines, and how they interact and behave in a complex setting. Fourier Analysis is simply a subtopic of a more general, and equally easy-to-understand subject, Linear Analysis. Linear Analysis is merely an extension of Linear Algebra, with all of the same rules and regulations, except that we are going to allow an infinite number of variables, or dimensions.

The basic concepts of Linear Analysis are the same as Linear Algebra. We start with the idea of a vector space, which involves objects and operations. There are two types of objects: (1) scalars, which are usually just real or complex numbers, and (2) vectors, which are just ordinary vectors in the Linear Algebra case and will often be sines and cosines of one type of another in Fourier Analysis. There then have to be two types of operations: (1) scalar multiplication, and (2) vector addition. Stated simply, if we have two scalars, c_1 and c_2, and two vectors, $\vec{v}_1$ and $\vec{v}_2$, then the combination $c_1\vec{v}_1 + c_2\vec{v}_2$ must also be a vector in our space. We generally insist that the scalars come from a field, which simply means they behave like the real or complex numbers.

A simple vector space would consist of all of the vectors of the type $c_1 \cos 2t + c_2 \sin 2t$. Addition and multiplication are well defined here, and there is a good reasons to consider these functions. We will revisit this vector space later.

We have not yet crossed into the world of Linear Algebra, but we have the basics in a vector space. If we are R^n, and we have a matrix, one of the basics of Linear Algebra is that if A is a matrix, then within the vector space, we have $A(c_1\vec{v}_1 + c_2\vec{v}_2) = c_1 A\vec{v}_1 + c_2 A\vec{v}_2$. Simply stated, the matrix respects the linear arithmetic of the vector space. You can distribute the multiplication by a matrix through scalar multiplication and addition. When we generalize this idea, they will be referred to not as matrices, but rather as linear operators.

1.3.1 The Dot Product, Inner Product, and Orthogonality

Length and Angle, Two Fundamental Quantities

The dot product is usually introduced in multivariable calculus and is extensively used in Linear Algebra. Simply stated, the dot product is defined by

$$\vec{a} \cdot \vec{b} = \sum a_k b_k.$$

This innocent-looking equation has far-reaching implications. First, it defines the standard notion of distance which we usually use, i.e.,

$$|\vec{a}|^2 = \sum |a_k|^2 = \vec{a} \cdot \vec{a}.$$

This can be seen to be an extension of the Pythagorean theorem. The sum of the squares of the sides is the length of the final vector squared. We define *length* as

$$|\vec{a}| = \sqrt{\vec{a} \cdot \vec{a}}.$$

The wonderful thing about the dot product is that it is not only valuable for determining length. The dot product also gives us a way to measure angles between vectors. The second important equation for the dot product is

$$\vec{a} \cdot \vec{b} = |\vec{a}||\vec{b}| \cos(\theta),$$

where θ is the angle between the vectors. By this equation, we implicitly define the angle between two vectors as

$$\cos(\theta) = \frac{\vec{a} \cdot \vec{b}}{|\vec{a}||\vec{b}|}.$$

Saying that we define this to be true is not accurate. This is the angle between any two vectors in $\mathbb{R}^3$.

In addition to defining length and angle, these equations give us one of our fundamental inequalities. Since $|\cos(\theta)| \leq 1$, the Cauchy–Schwartz inequality follows, stating that

$$|\vec{a} \cdot \vec{b}| \leq |a||b|.$$

For vectors in $\mathbb{R}^n$, imagining the angle between two vectors is a little more abstract. Recall, however, that even in $\mathbb{R}^3$, we define the angle between two vectors as the angle measured between them on the subspace created by the two vectors. Similarly, in $\mathbb{R}^n$ any two vectors define a two-dimensional subplane which is merely a copy of $\mathbb{R}^2$. Thus, the angle is also well defined in $\mathbb{R}^n$.

We want to extend these ideas of length and angle to functions. These functions will become the new vectors which we will analyze.

The Inner Product

We would like to consider functions on an arbitrary interval $[a, b]$. We would like to be able to measure their length and the angle between any two functions. An obvious first step is to pick points t_k uniformly spaced on the interval $[a, b]$, and consider the values of functions at these points. Suppose that we have two functions $f(t)$ and $g(t)$. We now consider the vectors $\vec{f} = \{f(t_k)\}$ and $\vec{g} = \{g(t_k)\}$. If there are N points, we now have approximations to the functions in $\mathbb{R}^N$. We can take the dot product of these approximations, and we get

$$\vec{f} \cdot \vec{g} = \sum_k f(t_k)g(t_k).$$

The problem with this dot product is that it is arbitrary, depending on how we pick the points t_k. We would like to think that if the t_k are close enough together, this dot product would mean something significant about the functions $f(t)$ and $g(t)$. If, however, we choose to have twice as many

points t_k, then the sum has twice as many elements and therefore will probably be nearly twice as big as it was before. Therefore, it is natural to change
this dot product to include the distance between points, so we consider

$$\vec{f} \cdot \vec{g} = \sum_k f(t_k)g(t_k)\delta_t. \tag{1.2}$$

We now ask the question, "Why not make the distance between points go
to zero?". At this point, we have the idea of an inner product.

Definition 1.3.1 (The Inner Product) *If $f(t)$ and $g(t)$ are two functions
on an interval $[a, b]$, we define their inner product to be*

$$\langle f(t), g(t) \rangle = \int_a^b f(t)g(t)dt. \tag{1.3}$$

Note that (1.2) is simply a Riemann sum. As long as the Riemann sum
converges, it converges to (1.3). We will avoid the specifics of when this
happens. We can state that the Riemann sum converges to the inner product
if $f(t)$ and $g(t)$ have only a finite number of discontinuities.

Let us summarize these two very important quantities and their properties.

Operation:	**The Dot Product:**	**The Inner Product:**
Computation:	$\vec{a} \cdot \vec{b} = \sum a_k b_k$	$\langle f(t), g(t) \rangle = \int f(t)g(t)dt$
Length:	$|\vec{a}|^2 = \vec{a} \cdot \vec{a}$	$\|f(t)\|^2 = \langle f(t), f(t) \rangle$
	$= \sum |a_k|^2$	$= \int |f(t)|^2 dt$
Angle:	$\vec{a} \cdot \vec{b} = |a||b|\cos(\theta)$	$\langle f(t), g(t) \rangle = \|f(t)\|\|g(t)\|\cos(\theta)$
Inequality:	$\vec{a} \cdot \vec{b} \leq |a||b|$	$\langle f(t), g(t) \rangle \leq \|f(t)\|\|g(t)\|$

Notice that we use simple absolute values for the length of a vector, $|\vec{a}|$, and
double these for a function, $\|f(t)\|$. The reason for this is unclear, but we will
stick with it, since it is standard.

Orthogonality

A fundamental idea of mathematics and geometry is that of the right
angle. It was certainly understood by the ancient Greeks, illustrated by the
Pythagorean theorem. When this idea is first introduced, it is generally stated
that two vectors or lines are perpendicular if the angle between them is 90
degrees. As we abstract this idea, mathematicians generally switch to the
word orthogonal. The origins of this are unknown to the author, but perpendicular and orthogonal imply the same thing.

Specifically, we will refer to two vectors or functions as being orthogonal if their dot products or inner products are 0. This can only happen if $\cos(\theta) = 0$, assuming the functions are nonzero. Moreover, we will say that a set of functions is orthogonal if the dot product or inner product of every subset of two of them is 0. Finally, we say that collections of vectors or functions are orthonormal if they are orthogonal, and each of them has length 1.

Realize that there is no difference between a function and a vector. A vector is a function on a discrete set of points. We can also think of functions of a continuous variable as vectors.

1.3.2 Eigenvectors and Eigenvalues in Linear Algebra

We will review some of the concepts of Linear Algebra which will be emphasized in this book. We do assume that the reader is familiar with Linear Algebra. Anyone not familiar with the basics is advised to refer to any one of the many excellent texts available on the subject.

Let us recall the concept of an eigenvector and its corresponding eigenvalue. Mathematically, this means that for an square $n \times n$ matrix A, $A\vec{v} = \lambda\vec{v}$ for some scalar λ. We refer to the eigenvalue λ and its corresponding eigenvector $\vec{v}$ as an eigenpair. This simply means that the direction of the vector is not changed, but its length is changed by the factor λ. Note that any multiple of an eigenvector is also an eigenvector. It is important to realize that if $\lambda < 0$, the vector points in the opposite direction, but is still parallel, or oppositely parallel to the original vector.

It is important to point out a number of common principles from Linear Algebra. The first is the idea of linear independence. A collection of vectors is said to be linearly independent if none of the individual vectors can be represented by a sum of the other vectors. A mathematical way to say this is that if $\vec{v}_1, \vec{v}_2, \ldots \vec{v}_k$ are linearly independent, then you can never express one of the vectors as the sum of the others, or you can never write $c_1\vec{v}_1 + c_2\vec{v} + c_3\vec{v}_3 \ldots c_k = 0$, unless all of the constants c_j are zero. This is because if one of the constants was nonzero, you could solve for that vector in terms of the other vectors, so they would not be independent.

There are a number of relationships between linear independence and eigenvalues/eigenvectors which must be understood. Any basic Linear Algebra book will go over these in detail. Here is a short list of some of the important ideas:

1. Every eigenvalue corresponds to at least one linearly independent eigenvector.

2. If two eigenvalues are not equal $\lambda_j \neq \lambda_k$, then the corresponding eigenvectors $\vec{v}_j$ and $\vec{v}_k$ are linearly independent.

3. If a collection of eigenvectors $\{\vec{v}_j\}$ has corresponding eigenvalues $\{\lambda_j\}$ which are all different, then the eigenvectors are all linearly independent.

4. An $n \times n$ matrix A has n eigenvalues, counting multiplicities.

5. If an $n \times n$ matrix A has n different eigenvalues, then it will have n eigenvectors which will be linearly independent.

This brings us to the fundamental idea of diagonalization. Let us suppose that we have a matrix A which has n linearly independent eigenvectors $\vec{v}_j$, with corresponding eigenvalues λ_j which we will assume are real for now. Then, for any n-dimensional vector $\vec{x}$, we can write

$$\vec{x} = c_1\vec{v}_1 + c_2\vec{v}_2 + c_3\vec{v}_3 \ldots c_n\vec{v}_n = \sum_{j=1}^{n} c_j\vec{v}_j. \qquad (1.4)$$

Now, let us consider multiplying the matrix A by $\vec{x}$. Using linearity $A\vec{x} = A(c_1\vec{v}_1 + c_2\vec{v}_2 + c_3\vec{v}_3 \ldots c_n\vec{v}_n) = c_1A\vec{v}_1 + c_2A\vec{v}_2 + c_3A\vec{v}_3 \ldots c_nA\vec{v}_n$ or in summation notation

$$A\vec{x} = A\left(\sum_{j=1}^{n} c_j\vec{v}_j\right) = \sum_{j=1}^{n} c_jA\vec{v}_j. \qquad (1.5)$$

The key here is that we do not actually have to multiply the matrix A by $\vec{x}$, or by any of the other vectors. We know that $A\vec{v}_j = \lambda_j\vec{v}_j$ which means that we have

$$A\vec{x} = \sum_{j=1}^{n} c_jA\vec{v}_j = \sum_{j=1}^{n} c_j\lambda_j\vec{v}_j. \qquad (1.6)$$

One might ask, "Why is this better?" It is better for a number of reasons.

1. We have reduced matrix multiplication $A\vec{x}$ into simple multiplication of scalars by vectors v_j.

2. Before we had an arbitrary large matrix, which is usually hard to interpret. Now, we have n individual eigenvectors, and we know exactly what happens to each of them. They either get expanded (if $|\lambda_j| > 1$) or shrunk (if $|\lambda_j| < 1$). Recall that if $\lambda_j < 0$, then the eigenvector points in the opposite direction.

3. In plain English, we have reduced a large confusing matrix to a bunch of one-dimensional eigenvector problems. We can easily visualize the one-dimensional eigenvector problems.

There are a number of other benefits to having the eigenvectors and eigenvalues. One example is that A will be invertible if and only if none of the λ_j are 0. Eigenvectors and eigenvalues allow one to visualize what is happening to individual vectors in a simple way, even when they come from a complex system.

1.3.3 Orthogonal Diagonalization

We want to consider an additional restriction to diagonalization. Let us suppose that not only does A have n linearly independent eigenvectors, but even more that A has orthogonal eigenvectors. By this, we mean that the dot product $\vec{v} \cdot \vec{v}_j = \vec{v}_i^t \vec{v}_j = 0$. We then store these eigenvectors as the columns of a matrix V and can normalize the entries of V that have length one, so $\vec{v}_i^t \vec{v}_i = 1$.

In this case, we then have $V^t V = I = V^{-1} V$ or more specifically $V^t = V^{-1}$. This allows us to write $A = V D V^{-1} = V D V^t$, and apply A to x in this situation. We get $A\vec{x} = V D V^t \vec{x}$ and $V^t \vec{x} = (\vec{v}_1 \cdot \vec{x}, \vec{v}_2 \cdot \vec{x}, \vec{v}_3 \cdot \vec{x}, \dots, \vec{v}_n \cdot \vec{x})^t$. Applying D to $V^t \vec{x}$ yields $D V^t \vec{x} = (\lambda_1(\vec{v}_1 \cdot \vec{x}), \lambda_2(\vec{v}_2 \cdot \vec{x}), \lambda_3(\vec{v}_3 \cdot \vec{x}), \dots, \lambda_n (\vec{v}_n \cdot \vec{x}))^t$. Finally, $A\vec{x} = V D V^t \vec{x} = \lambda_1(\vec{v}_1 \cdot \vec{x})\vec{v}_1 + \lambda_2(\vec{v}_2 \cdot \vec{x})\vec{v}_2 + \lambda_3(\vec{v}_3 \cdot \vec{x})\vec{v}_3 + \dots + \lambda_n(\vec{v}_n \cdot \vec{x})\vec{v}_n$. We can simplify the above into two simple sums. First, we can write

$$\vec{x} = \sum_{k=1}^{n} (\vec{v}_k \cdot \vec{x})\vec{v}_k.$$

More importantly, because of the nature of the eigenvectors, we can write

$$A\vec{x} = \sum_{k=1}^{n} \lambda_k (\vec{v}_k \cdot \vec{x})\vec{v}_k.$$

Thus, matrix multiplication can be reduced to a series of dot products with the eigenvectors, multiplied by the eigenvalues. This is a theme which we will utilize often in this book and which is often seen in science and engineering. This sometimes calls the eigenfunction approach.

1.3.4 Diagonalization in Linear Analysis: Eigenfunctions

We will now move forward and study eigenfunctions rather than just eigenvectors. Let us consider a simple system which is the model used for a violin or guitar string. This is illustrated in Figure 1.2. The string is held constant at both endpoints, but is pulled up before being released. The string acts very similarly to a spring, trying to move back to its resting or equilibrium place. The equilibrium is rather easily seen to be a simple line between the two fixed points. When it is released, it will move downward until it is once again a straight line. Instead of staying in this equilibrium position, however, the string's momentum will carry it beyond the equilibrium or zero level.

We will show later that a solution to this system is given by the series expansion

$$u(t, x) = \sum_{k=0}^{\infty} a_k \sin(kx) \cos(kt). \tag{1.7}$$

If we substitute the value $t = 0$ into this expression, we get the initial displacement of the system.

$$u(0, x) = f(x) = \sum_{k=0}^{\infty} a_k \sin(kx).$$

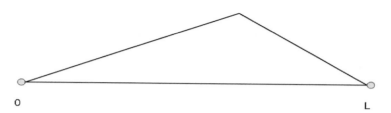

O   L

Figure 1.2: We illustrate a simple bound string problem above. The string is flexible and under tension.

The initial displacement $f(x)$ is represented in Figure 1.2. The coefficients a_k are therefore determined by the initial displacement. These coefficients and the general solution 1.7 determine the behavior of the system for the rest of time.

This might seem confusing, until you look at the individual terms in the series, or the functions

$$\sin(kx) \cos(kt).$$

Note that the frequency of the time variation, $\cos(kt)$, is tied to the frequency of the spatial variation $\sin(kx)$. This is the fundamental nature of what is known as the wave equation. We will study this equation in detail later. The wave equation models not only this simple string, but also the behavior of a drum and water waves, and many other phenomena.

We have illustrated the first three eigenfunctions of this simple string model in Figure 1.3. If we were to look at these three eigenfunctions evolve over time, the first would go through one cycle in the time period $[0, 2\pi]$. The second and third eigenfunctions, however, would go through two and three cycles, respectively.

1.3.5 Fourier Analysis

Fourier Analysis is Linear Analysis where we primarily study functions or phenomena which are easily described by sines and cosines. The list of topics here fills out most university catalogues. A small subset of topics would include the following: the study of light and optics, the study of sound, the study of sound waves in a medium, the study of light waves in medium, quantum mechanics, magnetic resonance imaging, computerized tomography, nuclear magnetic resonance, population dynamics, circuits and systems, digital communications, and imaging processing to name a few. Very few people ever master all of the topics above.

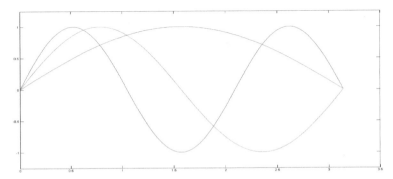

Figure 1.3:   We illustrate the first eigenfunction of a basic string above in blue, the second eigenfunction in red, and third eigenfunction in green. Musically, these are sometimes referred to as overtones. Note that in the time that the first eigenfunction completes one vibrational cycle, and the second and third eigenfunctions will complete two and three cycles, respectively.

The great thing about Fourier Analysis is that a thorough understanding of the topic will enable one to study all of the above at some level. Moreover, a thorough understanding of Fourier Analysis is required to understand any of the above topics.

1.3.6 Notational Differences

While Fourier Analysis is used widely across many disciplines, we rarely use the exact same notation. In this book, we will refer to the Fourier Transform of a function $f(t)$ as

$$\hat{f}(s) = \frac{1}{\sqrt{2\pi}} \int_{-\infty}^{\infty} f(t)e^{ist} dt.$$

In his very esteemed book [3], Ron Bracewell uses the Fourier Transform

$$\hat{f}(s) = \int_{-\infty}^{\infty} f(t)e^{-i2\pi st}dt.$$

In both cases, we are using s for the frequency component.

The difference for the notations is a matter of taste and represents a mere change of variables. The lack of a constant outside of Bracewell's integral is a positive for having the 2π in the integral. Oftentimes, s is used by electrical engineers, where ω is used by physicists. Mathematicians use a variety of variables, γ, f, etc. I find f difficult to use for frequency since I usually use it compulsively for functions. The important thing is to recognize that all of these are the same scientifically and mathematically, and to be able to read through the changes of notation.

Chapter 2
Basic Fourier Series

We will now study the basic rules of Fourier Series. You can study these basic principles for years in order to completely understand some of the many subtleties. The concentration of this book will be to quickly understand Fourier Series at a level which will allow to study their many applications. The reader is then encouraged to look further for a more in-depth understanding of the topics from the many other resources [1, 3, 5, 17, 20].

2.1 Fourier Series on $L^2[a, b]$

We begin with the most basic Fourier Series and outline how we can adjust them to other intervals. We will then outline the mathematics which forms the basis for these claims, and the many implications and structure which gives background to the study of Fourier Series.

We must begin with the basic definition of $L^2[a, b]$

Definition 2.1.1 *The set of functions $f(t) : [a, b] \to R$ whose squared integral is finite, or $\int_a^b |f(t)|^2 < \infty$ is referred to as $L^2[a, b]$, or the square integrable functions on $[a, b]$.*

Thus, the square integrable functions form the space of functions which we wish to study. We need to establish that if two functions $f(t), g(t) \in L^2[a, b]$, their linear sums $c_1 f(t) + c_2 g(t)$ are also in this space for any constants c_1 and c_2. To see this for real-valued functions, we simply write

---

Electronic supplementary material The online version of this chapter (https://doi.org/10.1007/978-1-4939-7393-4_2) contains supplementary material, which is available to authorized users.

$$\|c_1 f(t) + c_2 g(t)\|_2^2 = \int_a^b (c_1 f(t) + c_2 g(t))^2 dt$$

$$= \int_a^b c_1^2 f(t)^2 + c_1 c_2 f(t) g(t) + c_2^2 g(t)^2 dt$$

$$= c_1^2 \int_a^b f(t)^2 dt + c_2^2 \int_a^b g(t)^2$$

$$+ c_1 c_2 \int_a^b f(t) g(t) dt. \qquad (2.1)$$

The first and second integrals in (2.1) are finite because $f(t)$ and $g(t) \in L^2[a, b]$. The third integral is finite because of the Cauchy–Schwartz theorem which states that

$$\left| \int_a^b f(t) g(t) dt \right|^2 \le \|f(t)\|_2^2 \|g(t)\|_2^2.$$

Thus, it is a vector space. The proof when the functions are complex is left as an exercise.

We now state the most basic theorem of this section.

Theorem 2.1.1 *Let $f(t)$ be any function in $L^2[-\pi, \pi]$. Then, we can represent $f(t)$ in a series as*

$$f(t) = \frac{a_0}{2} + \sum_{k=1}^{\infty} a_k \cos(kt) + b_k \sin(kt)$$

where

$$a_k = \frac{1}{\pi} \int_{-\pi}^{\pi} f(t) \cos(kt) dt, \ \ and \ b_k = \frac{1}{\pi} \int_{-\pi}^{\pi} f(t) \sin(kt) dt.$$

Thus, you can represent any function in $L^2[-\pi, \pi]$ as a sum of sines and cosines. We can even state more than this:

Theorem 2.1.2 *Let $f(t)$ be any function in $L^2[a, b]$. Then, we can represent $f(t)$ in a series as*

$$f(t) = \frac{a_0}{2} + \sum_{k=1}^{\infty} a_k \cos\left(\frac{k\pi(t-h)}{H}\right) + b_k \sin\left(\frac{k\pi(t-h)}{H}\right)$$

where

$$a_k = \frac{1}{H} \int_a^b f(t) \cos\left(\frac{k\pi(t-h)}{H}\right) dt, \ and b_k = \frac{1}{H} \int_a^b f(t) \sin\left(\frac{k\pi(t-h)}{H}\right) dt.$$

The above constants are given by $H = (b - a)/2$, $h = (a + b)/2$.

Note that Theorem 2.1.2 is just a generalization of Theorem 2.1.1. To see this, let $a = -\pi$, $b = \pi$, which implies that $H = \pi$, and $h = 0$. Let us add a simplified version of Theorem 2.1.2, where the interval is centered about the origin, or where $a = -T$ and $b = T$.

Corollary 2.1.3 *Let $f(t)$ be any function in $L^2[-T, T]$. Then, we can represent $f(t)$ in a series as*

$$f(t) = \frac{a_0}{2} + \sum_{k=1}^{\infty} a_k \cos\left(\frac{k\pi t}{T}\right) + b_k \sin\left(\frac{k\pi t}{T}\right)$$

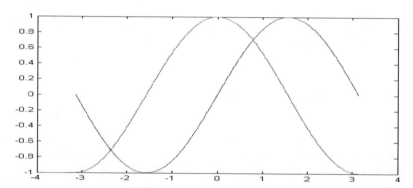

Figure 2.1:   Above you see that both the cosine (in green) and the sine (in blue) have exactly one cycle between $-\pi$ and π.

where

$$a_k = \frac{1}{T} \int_{-T}^{T} f(t) \cos\left(\frac{k\pi t}{T}\right) dt, \ and \ b_k = \frac{1}{T} \int_{-T}^{T} f(t) \sin\left(\frac{k\pi t}{T}\right) dt.$$

Theorem 2.1.1 is relatively easy to remember, while the author cannot remember Theorem 2.1.2 without recreating it. It is much easier to understand than it is to memorize, so let us understand how you can get Theorem 2.1.2 from Theorem 2.1.1 without memorization. To do this, we need pictures.

The key which is illustrated in Figure 2.1 is that the first cosine and the first sine in Theorem 2.1.1 (i.e., the cosine and sine terms with k = 1) have exactly one cycle between $-\pi$, and π. If we make the interval $[a, b] = [-2\pi, 2\pi]$, then we would have to change the cosine and sine terms to be $\cos(kt/2)$ and $\sin(kt/2)$. This assures that when t reaches the edge of the interval or 2π, we will have $kt/2 = k\pi$. This assures that the first cosine and sine terms of our new series (using $\cos(kt/2)$ with $k = 1$) will also have exactly one cycle.

To understand why this works, imagine a function $f(t)$ expanded on $[-\pi, \pi]$. Now, stretch that function to $[-2\pi, 2\pi]$ by considering $f(t/2)$. Notice that $t/2$ travels from $-\pi$ to π in the time that t goes from -2π to 2π. All we did by substituting $t/2$ into the cosine or sine terms is adjust the series that we had formerly expanded, or if

$$f(t) = \frac{a_0}{2} + \sum_{k=1}^{\infty} a_k \cos(kt) + b_k \sin(kt)$$

then surely

$$f(t/2) = \frac{a_0}{2} + \sum_{k=1}^{\infty} a_k \cos(kt/2) + b_k \sin(kt/2).$$

The one detail is that we do have to calculate the coefficients a_k and b_k slightly differently on our new interval. You can, however, do a change of variable in the old formula and get the same thing.

2.1.1   Calculating a Fourier Series

The obvious problem with Theorem 2.1.1 is that we do not just have to calculate one integral, but an infinite number of them. The reality is that oftentimes we can calculate all of these integrals simultaneously. The second reality is that except for a certain number of simple functions, the Fourier Series cannot easily be calculated by hand. The Fourier Series will be evaluated by the computer in general, but that is a topic for a later chapter.

Example 1:

Let us begin with one of the most basic functions for which Fourier Series is used. This is a common function which is oftentimes used to model an on–off switch in electrical engineering. Let us consider what we will call the characteristic function on $[-\pi, \pi]$ which we define to be

$$\chi(t) = \begin{cases} 1 \text{ if } |t| < \frac{\pi}{2} \\ 0 \text{ if } |t| > \frac{\pi}{2} \end{cases}. \tag{2.2}$$

We will use this function often so we will introduce the notation

$$\chi_a(t) = \begin{cases} 1 \text{ if } |t| < a \\ 0 \text{ if } |t| > a \end{cases}, \tag{2.3}$$

or

$$\chi_{[a,b]}(t) = \begin{cases} 1 \text{ if } |t| \in [a, b] \\ 0 \text{ if } |t| \notin [a, b] \end{cases}. \tag{2.4}$$

Now, we need to calculate the coefficients

$$a_k = \frac{1}{\pi} \int_{-\pi}^{\pi} f(t) \cos(kt) dt, \text{ and } b_k = \frac{1}{\pi} \int_{-\pi}^{\pi} f(t) \sin(kt) dt,$$

from Theorem 2.1.1. There are a couple of things we need to remember from basic calculus to make our lives easier. The first is the definition of even and odd functions.

Definition 2.1.2 (Even and Odd Functions) *A function $f(t)$ is said to be even if $f(t) = f(-t)$. A function $f(t)$ is said to be odd if $f(t) = -f(-t)$.*

The first obvious reason why we care about even and odd functions is that $\cos(kt)$ is even for all k and $\sin(kt)$ is odd for all k. Another set of functions which separate nicely into even and odd functions is the monomials t^k. If k is even, t^k is even. If k is odd, t^k is odd.

To understand why this helps, remember that if $f(t)$ is odd, $\int_{-T}^{T} f(t) dt = 0$. Now, remember that the product of two even functions is even. The product of an even function and an odd function is odd. Finally, the product of an odd function with an odd function is even. These correspond directly to the products of positive and negative numbers, where even is positive and odd is negative, for some rather obvious reasons.

Now, consider the formulas for a_k and b_k above. If $f(t)$ is even, then $f(t) \sin(kt)$ is odd, so the b_k terms will all be zero. This makes sense because an even function can be represented entirely by cosine (even) terms. Similarly, if $f(t)$ is odd, $f(t) \cos(kt)$ will be odd, so the a_k terms are all zero. Thus, the Fourier Series also separates our functions nicely into even and odd terms.

Let us return to calculating the Fourier Series for $\chi(t)$ as defined above. Notice that $\chi(t)$ is even, so as a result all of the sine terms will be zero. Let us now consider the a_k or cosine terms. To begin with this seems difficult, but the key is to separate the integral into the two parts, namely when $|t| < \frac{\pi}{2}$ and when $|t| > \frac{\pi}{2}$. Mathematically, this becomes

$$a_k = \frac{1}{\pi} \int_{-\pi}^{\pi} \chi(t) \cos(kt) dt = \frac{1}{\pi} \int_{-\pi/2}^{\pi/2} \chi(t) \cos(kt) dt + \frac{1}{\pi} \int_{\pi/2 < |t| \leq \pi} \chi(t) \cos(kt) dt.$$

Note that $\chi(t) = 1$ in the first integral and $\chi(t) = 0$ in the second integral is zero so we have

$$a_k = \frac{1}{\pi} \int_{-\pi/2}^{\pi/2} 1 \cos(kt) dt + 0 = \frac{1}{\pi} \frac{\sin(kt)}{k} \Big|_{-\pi/2}^{\pi/2}, \tag{2.5}$$

$$= \frac{1}{\pi} \frac{\sin(k\pi/2) - \sin(-k\pi/2)}{k} = \frac{2 \sin(k\pi/2)}{k\pi}. \tag{2.6}$$

Note that we used the odd property of the sine function to finish the last step, namely $\sin(x) - \sin(-x) = 2 \sin(x)$. This is a fine and complete answer which will allow us to calculate the series without difficulty. Further examination, however, shows that when k is even, the terms are zero, i.e., $\sin(m\pi) = 0$ for all integers m. Examining once again shows that for k odd, $\sin(k\pi/2) = (-1)^{(k+1)/2}$. Thus, it is 1 for $k = 1$ and -1 for $k = 3$, and it alternates through the series. This is a nice formula, but notice that (2.5) is not valid for $k = 0$, since $\cos(0t) = 1$, and $\int 1dt = t$. This individual case is easily solved, and we get $\frac{a_0}{2} = \frac{1}{2}$. This is also the average value of $\chi(t)$ which will always be the case, from the definition of $\frac{a_0}{2}$. Thus, the series is

$$\chi(t) = \frac{1}{2} + \frac{2}{\pi}\cos(t) - \frac{2}{3\pi}\cos(3t) + \frac{2}{5\pi}\cos(5t) - \frac{2}{7\pi}\cos(7t)\ldots . \quad (2.7)$$

While the expansion formula (2.7) is nice, it is actually harder to program on a computer than the most basic formula which we derived in (2.5). This formula is then

$$\chi(t) = \frac{1}{2} + \sum_{k=0}^{\infty} \frac{2\sin(k\pi/2)}{k\pi}\cos(kt).$$

Note that the above formula has every other term being zero, but it is actually easier to program on a computer. Thus, sometimes trying to reduce the formulas too far actually requires more work later.

We are now going to try to see what this series, or the partial sums of this series look like. No one ever evaluates an infinite sum unless it is a very special sum. By partial sums, we are referring to the approximations

$$S_n(\chi(t)) = \frac{1}{2} + \sum_{k=0}^{n} \frac{2\sin(k\pi/2)}{k\pi}\cos(kt). \quad (2.8)$$

A crucial question in Fourier Analysis is "How quickly do these sums begin to approximate $\chi(t)$?" These approximations are shown in Figure 2.2. You can see from this figure that the basic shape of the function is starting to approximate the function. Note that these sums are only guaranteed to approximate the function according to the average error metric $E_n(\chi) = \int_{-\pi}^{\pi} |f(t) - S_n(t)|^2 dt$. A quick examination of the numerics used to generate the plots shows that the errors $E_n(\chi)$ are 1.57 for $n = 0$, and 0.29, 0.15, and 0.10 for $n = 1, 3, 5$, respectively. Remember that every other term in the approximation is zero, so it does not get better for $n = 2, 4, 6$ etc.

To investigate further we check the numerical approximation of $\chi(t)$ with 20, 50, and 100 terms. The numerical results for the error are 0.03, 0.01, and 0.008. Thus, the error is going to zero, although very slowly. These results are shown in Figure 2.3. We will spend extensive time in the future understanding the rate at which these series converge, and the many implications of these convergence rates.

Example 2:

Before we move on to the next topic, we will address a second example, point-ing the way to the solution which is left as an exercise. Let us represent the extremely simple function $f(t) = t$ in terms of its Fourier Series, on the sim-ple interval from $[-\pi, \pi]$. Let us explain why we call this the simple interval. This is because you do not need anything inside $\cos(kt)$ or $\sin(kt)$. It can be argued that $[-1, 1]$ is a simpler interval, but then you have $\cos(k\pi t)$, etc. Neither is very difficult.

The coefficients for the function $f(t)$ will be exclusively sine coefficients, since t is an odd function. Recall odd functions only have sine coefficients. If you get confused, just compute them all, and hopefully, the cosine coefficients will be zero. It is good to understand, however, since it saves time and allows one to correct arithmetic mistakes. So our sine coefficients will be

$$b_k = \frac{1}{\pi} \int_{-\pi}^{\pi} t \sin(kt) dt.$$

This evaluation is trickier than the prior example. One must remember inte-gration by parts from first-year calculus. This is not all that time consuming and is left as an exercise. We do illustrate the approximations in Figure 2.4.

It is informative that if we consider $f(t) = t^2$, then we have to do inte-gration by parts twice and three times for $f(t) = t^3$. This brings up one of the problems with basic Fourier Series. Calculating the integral coefficients by hand

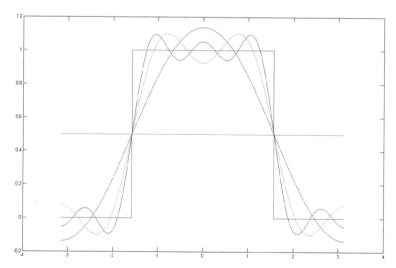

Figure 2.2: We have plotted the first four nonzero approximations of the Fourier expansion derived above in (2.7). You can see the terms start to approximate the function, although they are not exact with only n = 1, 3, and 5 terms, respectively.

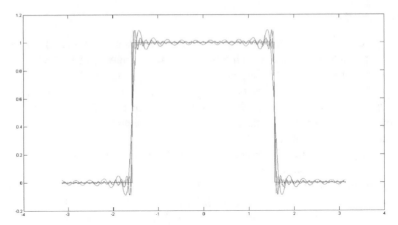

Figure 2.3: We have plotted the first approximations to $\chi(t)$ above, with 20, 50, and 100 terms. The terms are starting to approximate the function very closely, but there is still a lot of "ringing" in the approximation which is not present in the function.

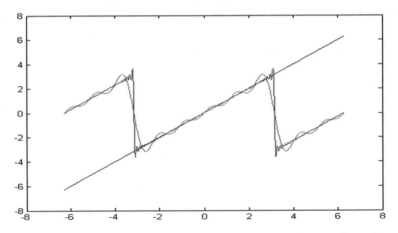

Figure 2.4: We have plotted the first approximations to $f(t) = t$ on $[-\pi, \pi]$ above, with 5 and 50 terms. The terms are starting to approximate the function very closely, but there is still a lot of "ringing" in the approximation which is not present in the function. Note also that the approximations are only valid on $[-\pi, \pi]$, and the representation is periodic and not at all valid outside of $[-\pi, \pi]$.

is not easy for most functions and impossible for some. As a result, we will study methods to calculate these coefficients by other means, most often with numerical approximations.

2.1.2   Periodicity and Equality

Fourier Series can approximate nearly any function over a finite interval. One should not make the mistake of believing that the Fourier Series will somehow represent the function outside of that predesigned interval. Recall that if the interval is $[-\pi, \pi]$, then the Fourier Series takes on the form

$$f(t) = \frac{a_0}{2} + \sum_{k=1}^{\infty} a_k \cos(kt) + b_k \sin(kt). \qquad (2.9)$$

All of the functions in the above expansion are 2π periodic. Specifically, $\cos(kt) = \cos(k(t + m2\pi))$ and $\sin(kt) = \sin(k(t + n2\pi))$. Therefore, we must consider two different things, the function $f(t)$ and the Fourier expansion of $f(t)$, which we will define to be

$$S(f)(t) = \frac{a_0}{2} + \sum_{k=1}^{\infty} a_k \cos(kt) + b_k \sin(kt).$$

Since $S(f)(t)$ consists entirely of 2π periodic functions, it follows that $S(f)(t) = S(f)(t + m2\pi)$ and is also 2π periodic.

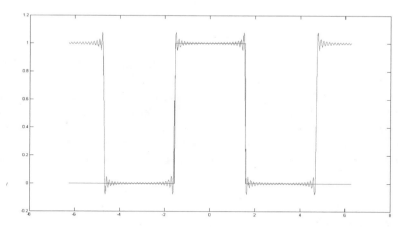

Figure 2.5: We have plotted an approximation to $\chi(t)$ above, with 50 terms on $[-2\pi, 2\pi]$. Note that this approximation is relatively good from $[-\pi, \pi]$, but does not approximate $\chi(t)$ outside of the interval. Instead it repeats itself, as we have stated.

For this reason, we can only say that $f(t) = S(f)$ on the interval $[-\pi, \pi]$. This is completely reasonable, because the coefficients a_k and b_k only depend on $f(t)$ in that interval. Secondly, remember that $f(t)$ does not need to be periodic at all, and so it is obvious that we cannot expect that $S(f)(t)$ would represent $f(t)$ outside of the defined interval. Perhaps we should refer to $S(f)(t)$ as $S_\pi(f)(t)$, meaning that it is the Fourier Series on $[-\pi, \pi]$, and $S_{[a,b]}(f)(t)$ to be the Fourier Series on $[a,b]$. When it is important to make the distinction, we will use this altered notation.

To emphasize this point, we should look at $S_\pi(f)(t)$, or one of its approximations, on $[-2\pi, 2\pi]$. This is shown in Figure 2.5.

At this time, we need to address the issue of how we are using equality Theorem 2.1.1, and in all of the stated equalities in this chapter. What does it mean that a function

$$f(t) = \frac{a_0}{2} + \sum_{k=1}^{\infty} a_k \cos(kt) + b_k \sin(kt)?$$

The right-hand side is an infinite sum, so what we would like to say is that the limit as the number of terms goes to ∞ converges to $f(t)$. Even this is not the correct statement. Let us define

$$S_n(f) = \frac{a_0}{2} + \sum_{k=1}^{n} a_k \cos(kt) + b_k \sin(kt).$$

What we really mean is that the error function defined as

$$E_n(f) = \int_{-\pi}^{\pi} |f(t) - S_n(t)|^2 dt$$

goes to zero as $n \to \infty$. Specifically, we know that $\lim_{n\to\infty} E_n(f(t)) = 0$ for any function $f(t) \in L^2[a,b]$. We cannot guarantee that the same is true if we remove the integral in the definition of $E_n(f)$. Specifically, we will not have $|f(t) - S_n(t)| \to 0$ for all $f(t) \in L^2(a,b)$. Pointwise convergence in $L^2[a,b]$ does not really make sense however, since we say that two functions are equal if $\int |f(t) - g(t)|^2 dt = 0$. Remember that integrals do not take into account the value of functions at single points. In general, the partial sums do approach the function $f(t)$ at each specific point, at least with the exception of a set of measure zero, which is beyond the scope of this course. This was proven relatively recently [4].

There are many different books which discuss the relative convergence rates of Fourier Series. These are important and interesting, and the reader is encouraged to look into these topics [1, 3, 17, 20]. They are beyond the scope of this book, however. We will consider only Fourier Series in L^2 of some type or another. We will then use these convergence rates to understand some of the subtleties of the applications.

2.1.3 Problems and Exercises

1. Compute the Fourier Series for the function

$$\chi_{\pi/4}(t) = \begin{cases} 1 \text{ if } |t| < \frac{\pi}{4} \\ 0 \text{ if } |t| > \frac{\pi}{4} \end{cases}, \qquad (2.10)$$

on $[-\pi, \pi]$ by using the above example as a guide. Also, plot the first few terms of the expansion (a) on $[-\pi, \pi]$ and (b) on $[-2\pi, 2\pi]$.

2. Compute the Fourier Series for $\chi_{\pi/4}(t)$ on $[-2\pi, 2\pi]$. Plot this on the interval from $[-6\pi, 6\pi]$ or a larger interval and compare to the plot on the function expanded on $[-\pi, \pi]$. Realize that is should converge on the whole interval $[-2\pi, 2\pi]$, and you must use the appropriate sine and cosine terms $\cos(k/2t)$ and $\sin(k/2t)$.

3. Compute two Fourier Series for the function $f(t) = t$ (a) on $[-\pi, \pi]$ and (b) on $[-2\pi, 2\pi]$. Plot the approximations using 5,10, and 15 terms on $[-4\pi, 4\pi]$.

4. (a) Find the necessary cosine and sine terms for expanding on $[-1, 1]$ and the appropriate coefficient formulas to approximate a function on $[-1, 1]$. (b) Compute the expansion of $f(t) = t^2$ on $[-1, 1]$, and plot the first few terms on $[-3, 3]$.

5. (a) Figure out the necessary cosine and sine terms and the appropriate coefficient formulas to approximate a function on the interval $[1, 3]$. (b) Using the result from 4 above, plot the first few terms of the expansion of t on $[1, 3]$ and on $[-1, 5]$.

6. Find the expansion for $f(t) = t^3$ on $[-1, 1]$. Plot the first few terms on the interval $[-3, 3]$.

7. Compute the expansion of $f(t) = \cos(t/2)$ on $[-\pi, \pi]$. Plot the first few terms on this interval.

8. Compute the expansion of $f(t) = \cos^2(t/2)$ on $[-\pi, \pi]$. Plot the first few terms on this interval.

9. Compute the expansion of $f(t) = \sin(t/2)$ on $[-\pi, \pi]$. Plot the first few terms on this interval.

10. Compute the expansion of

$$f(t) = \begin{cases} 0 & \text{for } |t| > \pi/2 \\ t + \pi/2 & \text{for } t \in [-\pi/2, 0] \\ \pi/2 - t & \text{for } t \in [0, \pi/2] \end{cases} \qquad (2.11)$$

on the interval $[-\pi, \pi]$. Plot the first few terms on this interval.

11. Compute the expansion of

$$f(t) = \begin{cases} 0 & \text{for } |t| > \pi/2 \\ -1 & \text{for } t \in [-\pi/2, 0] \\ 1 & \text{for } t \in [0, \pi/2] \end{cases} \qquad (2.12)$$

on the interval $[-\pi, \pi]$. Plot the first few terms on this interval.

12. Compute the expansion of

$$f(t) = \begin{cases} 0 & \text{for } |t| < 0 \\ 1 & \text{for } |t| \geq 0 \end{cases} \qquad (2.13)$$

on the interval $[-\pi, \pi]$. Plot the first few terms on this interval. Plot them on $[-2\pi, 2\pi]$ also.

2.2 Orthogonality and Hilbert Spaces

We simply stated the basics of Fourier Series in the last section. We intentionally presented no proofs, and we also skipped what we will refer to as the geometry of Fourier Series. We will refer to the discussion of orthogonality which we touched on in the first chapter. Orthogonality generally comes up in Linear Algebra. Its extension to studying functions here is straightforward. We will review quickly the basic concepts. The first and most obvious statement or question is "What is the difference between orthogonality and perpendicularity?" The answer is there is no difference. They are interchangeable, but mathematicians and most others generally switch to orthogonality when dealing with higher dimensional vector spaces, i.e., $\mathbb{R}^n$, where $n > 3$, or function spaces, such as $L^2[a, b]$ as we defined in the last section.

Let us first begin with some notation. From this point forward, we will cease to recognize the difference between a common dot product, as we have in basic Linear Algebra, and the inner product, which we have yet to define. The dot product of two vectors is given by $\vec{a} \cdot \vec{b} = \vec{a}^t b = \sum_k a_k b_k$. Namely, we multiply and add. Remember that length (common Euclidean distance) is given by $|\vec{a}| = \sqrt{\vec{a} \cdot \vec{a}}$. Recall also that the angle θ between two vectors is given by $\vec{a} \cdot \vec{b} = |a||b| \cos(\theta)$.

To define the inner product as we will use it in this book, think first of how we would compare two functions $f(t)$ and $g(t)$. We might start by approximating length by using the dot product, perhaps sampling (or discretizing) the functions to a finite number of points t_k, where $t_0 = a$ and $t_n = b$ on an interval $[a, b]$. It follows that the distance between consecutive points would be $\delta t = (b - a)/n$. Thus, we would like to approximate the length of these functions by their vector counterparts, or say that $f(t) \cdot g(t) \approx \sum_k f(t_k) g(t_k)$. The one problem is that if we use twice as many points, the length will approximately double. This is recognized by multiplying by the distance between points or using

$$f(t) \cdot g(t) \approx \sum_k f(t_k)g(t_k)\delta t.$$

We recognize the above from first-year calculus as a Riemann sum. The Fundamental Theorem of Calculus then tells us if we let δt go to zero, the limit should converge (in most cases), and therefore, we have our definition of an inner product

Definition 2.2.1 (Inner Product) *If $f(t)$ and $g(t)$ are functions in $L^2[a,b]$, then we define the inner product between $f(t)$ and $g(t)$ to be*

$$(f(t) \cdot g(t)) \equiv \langle f(t), g(t) \rangle = \int_a^b f(t)\overline{g(t)}dt.$$

Moreover, the length of $f(t)$ is designated to be $\|f(t)\| = \sqrt{(\int_a^b |f(t)|^2 dt} = \sqrt{\langle f, f \rangle}$.

Note that we had to use the complex conjugate in the inner product definition above. This allows us to deal with complex-valued functions. If we did not use this, note that a function which is purely complex would have a negative distance, which we do not want. We will oftentimes not include the t in the integral and just write $\langle f, g \rangle$ for the inner product.

Any standard text on linear analysis or Linear Algebra will go into the details of inner products. We encourage the reader to check out these more detailed expositions. The important part here is that $L^2[a,b]$ will inherit all of the properties that we were used to in Linear Algebra, with the standard dot product. We state the important ones here:

1. (Linearity) If f and g are any two functions in an inner product space, then $c_1 f + c_2 g$ is also a function in that inner product space.

2. If $f \neq 0$, then $\|f\| \neq 0$.

3. (Triangle Rule) If f and g are in an inner product space, $\|f + g\| \leq \|f\| + \|g\|$.

4. (Cauchy–Schwartz Inequality) If f and g are in an inner product space

$$|\langle f, g \rangle| \leq \|f\|\|g\|. \tag{2.14}$$

Moreover, equality holds only if f is a constant multiple of g, or $f(t) = cg(t)$.

Let us recall that the Cauchy–Schwartz inequality is an outgrowth of the standard formula in Linear Algebra $\vec{a} \cdot \vec{b} = |a||b| \cos(\theta)$. Since $\cos(\theta)$ is always less than or equal to one, the inequality holds.

We will now focus on orthogonality, since it is a key which we must keep in mind. Simple functions such as the monomials $1, t, t^2, t^3 \ldots$ are in L^2 of any interval, but they are not even close to being orthogonal. Let us now formalize this idea which was introduced in Chapter 1.

Definition 2.2.2 *Two functions $f(t)$ and $g(t)$ in $L^2[a,b]$ are said to be orthogonal if $\langle f, g \rangle = 0$. A collection or set of functions $\{o_k(t)\}_{k=0}^N$ is said to be orthogonal if for any pair functions from the set $\langle o_j, o_i \rangle = 0$ as long as $i \neq j$. In addition, we can say that a set of functions is orthonormal if they are orthogonal, and they all have length one, or $\|o_j\| = 1$.*

One of the keys to utilizing Fourier Series is that sines and cosines, when adjusted for an interval as in Theorem 2.1.1 and Theorem 2.1.2, are naturally orthogonal. This makes the computation used in Theorem 2.1.1 work. Let us first state the result.

Theorem 2.2.1 (Orthogonality) *The functions $\{\cos(kt)\}_{k=0}^\infty$, and $\{\sin(kt)\}_{k=1}^\infty$ are orthogonal on the interval $[-\pi, \pi]$. In addition, the functions used in Theorem 2.1.2 are also orthogonal on the corresponding interval $[a, b]$.*

To prove this, we will need to remember the trigonometric addition identities: $\sin(\alpha \pm \beta) = \sin(\alpha)\sin(\beta) \pm \cos(\alpha)\cos(\beta)$ and $\cos(\alpha \pm \beta) = \cos(\alpha)\cos(\beta) \mp \sin(\alpha)\sin(\beta)$.

Proof: We will proceed to prove three things. (1) All of the cosines are orthogonal to all of the sines, and (2) The cosines are orthogonal to each other, and (3) The sines are orthogonal to each other.
Proof of (1): This is easy, since cosine is even and sine is odd, which implies that $\cos(mt)\sin(nt)$ is odd, so it follows immediately that

$$\int_{-\pi}^{\pi} \cos(mt)\sin(nt) = 0$$

regardless of m or n. This is simple because the integral of an odd function on a symmetric interval is zero, since the right and left halves of the integral are negatives of one another.
Proof of (2): This requires the double angle trigonometric identities. We are considering $\cos(mt)\cos(nt)$. If we look at the identities, we have that $\cos(mt + nt) = \cos(mt)\cos(nt) - \sin(mt)\sin(nt)$ and $\cos(mt - nt) = \cos(mt)\cos(nt) + \sin(mt)\sin(nt)$. Adding these identities yields

$$\cos(mt + nt) + \cos(mt - nt) = 2\cos(mt)\cos(nt).$$

Now substituting, we have

$$\int_{-\pi}^{\pi} \cos(mt)\cos(nt)dt = \int_{-\pi}^{\pi} \frac{1}{2}(\cos(mt+nt)+\cos(mt-nt))dt$$

$$= \frac{1}{2}\left[\frac{\sin((m+n)t)}{m+n} + \frac{\sin((m-n)t)}{m-n}\right]_{-\pi}^{\pi}.$$

Note that we have assumed in the last integration step that $m \neq n$. Now, recall that $\sin(k\pi) = 0$ whenever k is an integer. Thus, both terms on the right are zero. Notice also that if $m = n$, integration of the $\cos(mt - nt) = 1$ term on the right yields the squared length of $\cos(mt)$, which is π. This yields the $\frac{1}{\pi}$ term in Theorem 2.1.1.

Proof of (3): This is identical to the Proof of (2) with the exception of using $\cos(mt+nt) - \cos(mt-nt)$ to cancel the cosine terms on the right.

We have proven orthogonality of the functions in Theorem 2.1.1. The proof for the functions in Theorem 2.1.2 is identical as long as you note the following things. The functions in Theorem 2.1.2 are adjusted so that just in the Proof of part (2) above, the sine terms will be zero at the endpoints. This is due to the fact the sine terms are adjusted to have exactly one cycle for $k = 1$ and multiple cycles $k > 1$ (k is an integer).

2.2.1   Orthogonal Expansions

We have avoided discussions of completeness, or whether or not we have enough functions for the expansions which we are suggesting. An interested student should inquire of this, but it is beyond the scope of this book.

We will prove Theorems 2.1.1 and 2.1.2 given the completeness of those functions. This is an easy process. By completeness, we mean that for any function $f(t) \in L^2[-\pi, \pi]$, we know that there is an expansion of $f(t)$ such that

$$f(t) = \frac{a_0}{2} + \sum_{k=0}^{\infty} a_k\cos(kt) + b_k\sin(kt).$$

We will now show that the values for a_k and b_k as stated in Theorems 2.1.1 and 2.1.2 are correct. We know that

$$\langle f(t), \cos(jt) \rangle = \int_{-\pi}^{\pi} f(t)\cos(jt)dt$$

$$= \int_{-\pi}^{\pi} \left(\frac{a_0}{2} + \sum_{k=1}^{\infty} a_k\cos(kt) + b_k\sin(kt)\right)\cos(jt)dt$$

$$= \int_{-\pi}^{\pi} \frac{a_0}{2}\cos(jt)dt + \sum_{k=1}^{\infty} \int_{-\pi}^{\pi} a_k\cos(kt)\cos(jt)dt + \int_{-\pi}^{\pi} b_k\sin(kt)\cos(jt)dt.$$

Note that all of the terms on the right are 0 except the cosine term when $k = j$, so we have

$$a_k \int_{-\pi}^{\pi} \cos^2(kt)dt = \int_{-\pi}^{\pi} f(t)\cos(kt)dt.$$

Remembering that we proved that right-hand term to be π in the orthogonality section, we have that

$$a_k = \frac{1}{\pi} \int_{-\pi}^{\pi} f(t)\cos(kt)dt.$$

Now, we will adjust our notation slightly, so that we can deal with orthonormality. When we deal with the sine and cosine terms $\sin(kt)$ and $\cos(kt)$ on $[-\pi, \pi]$, their squared lengths in $L^2[a, b]$ are π as stated above. So to get an *orthonormal* set of functions, we have to consider $\frac{1}{\sqrt{\pi}}\sin(kt)$ and $\frac{1}{\sqrt{\pi}}\cos(kt)$. In addition, the constant term 1 has a length of $\sqrt{2\pi}$ so its orthonormal version is $\frac{1}{\sqrt{2\pi}}$. Thus, the orthonormal version of the Fourier Series is

$$f(t) = \frac{a_0}{\sqrt{2\pi}} + \sum_{k=1}^{\infty} a_k \frac{\cos(kt)}{\sqrt{\pi}} + b_k \frac{\sin(kt)}{\sqrt{\pi}}, \tag{2.15}$$

where

$$a_k = \int_{-\pi}^{\pi} \frac{\cos(kt)}{\sqrt{\pi}} f(t)dt, \text{ and } b_k = \int_{-\pi}^{\pi} \frac{\sin(kt)}{\sqrt{\pi}} f(t)dt,$$

for $k \geq 0$ and

$$a_0 = \int_{-\pi}^{\pi} \frac{1}{\sqrt{2\pi}} f(t)dt.$$

If we compare the representation in (2.15) to that in (2.1.1), we see that with the exception of the placement of constants, they are identical. The change in notation will be necessary in order to realize the benefits of the orthonormal expansion (2.15).

If we change this to an expansion on $[-T, T]$, then we have

$$f(t) = \frac{a_0}{\sqrt{2T}} + \sum_{k=1}^{\infty} a_k \frac{\cos(k\pi t/T)}{\sqrt{T}} + b_k \frac{\sin(k\pi t/T)}{\sqrt{T}}, \tag{2.16}$$

where

$$a_k = \int_{-T}^{T} \frac{\cos(k\pi t/T)}{\sqrt{T}} f(t)dt, \text{ and } b_k = \int_{-T}^{T} \frac{\sin(k\pi t/T)}{\sqrt{T}} f(t)dt,$$

for $k \geq 0$ and

$$a_0 = \int_{-T}^{T} \frac{1}{\sqrt{2T}} f(t)dt.$$

More generally, we have the following theorem, whose proof is identical to that above.

Theorem 2.2.2 *Let* $\{o_k\}$ *be a complete set of orthonormal functions in* $L^2[a, b]$ *then for any function in* $L^2[a, b]$, *we have*

$$f(t) = \sum_k \langle f(t), o_k \rangle o_k.$$

This theorem is true in any inner product space, but we state it for our current setting. We will emphasize further generalizations when relevant. From now on, we will refer to any topologically complete inner product space as a Hilbert space. We will not go into the details of topological completeness. It suffices to state that all inner product spaces in this book are Hilbert spaces.

2.2.2   Problems and Exercises:

1. Prove that the representation in (2.15) is correct. Specifically, show that the norm, or length of the functions $\cos(kt)$ and $\sin(kt)$ is π on $[-\pi, \pi]$, and that the length of 1 is 2π on this interval.

2. **Challenging:** Rewrite the expansions in Theorem 2.1.2 so that it is an orthonormal expansion such as in 2.15. Show that these functions are orthogonal on $[a, b]$. Find the norms of these functions on $[a, b]$, and give the altered version of Theorem 2.1.2. This theorem should distribute the normalization factors as in (2.15).

2.3   The Pythagorean Theorem

One of the oldest and most well-known theorems for elementary students and others dates back to the Greeks. Namely, in a right triangle, or a triangle in which one of the angles is $90°$, the squared sum of the length of two sides is the square of the third side, or $a^2 + b^2 = c^2$. This simple elementary school identity has fundamental importance in Fourier Analysis and Linear Analysis in general. It is also very simple to prove.

Theorem 2.3.1 (Pythagorean Theorem) *Let* $f(t)$ *be a function in a Hilbert space* H *(such as* $L^2[a, b]$*). Let* $\{o_k\}_k$ *be a complete set of orthonormal functions and* $f(t)$ *be an element of* H. *We know from Theorem 2.2.2 that*

$$f(t) = \sum_k \langle f(t), o_k \rangle o_k.$$

The Pythagorean theorem states that in addition, the squared length of f, i.e.,
$\|f\|^2$, *can be represented as*

$$\|f\|^2 \equiv \langle f, f \rangle = \sum_k |\langle f, o_k \rangle|^2.$$

In addition, we have that if $f = \sum_k a_k o_k$ and $g = \sum_k b_k o_k$, then

$$\langle f, g \rangle = \sum_k a_k b_k.$$

Stated simply, the squared length of f is identical to the sum of the squared
lengths of the sides, or $\sum_k |\langle f, o_k \rangle|^2$. This is exactly the Pythagorean theorem
which is taught at the elementary school. The interesting thing is that the
proof can be understood in middle school (perhaps), but it is exactly the
same in this highly abstract setting.

Proof: Since we know f is in H, we simply begin with the orthogonal decomposition in Theorem 2.2.2

$$f(t) = \sum_k \langle f(t), o_k \rangle o_k = \sum_k c_k o_k.$$

where for simplicity of notation we are letting $c_j = \langle f(t), o_j \rangle$. Now, by the
definition of distance or length, we have

$$\|f\|^2 = \langle f, f \rangle = \left\langle \sum_j a_j o_j, \sum_k a_k o_k \right\rangle$$
$$= \sum_j a_j \left\langle o_j, \sum_k a_k o_k \right\rangle.$$

In the second line above, we have moved the sum and the constants a_j outside
of the inner product. This is equivalent to moving the sum and constants
outside of the integral, which is permitted by the linearity of the integral. We
can do this once again with the second sum, and we get

$$\|f\|^2 = \sum_j a_j \left\langle o_j, \sum_k a_k o_k \right\rangle$$
$$= \sum_j \sum_k a_j a_k \langle o_j, o_k \rangle.$$

Now, recall that the $\{o_j\}$ are an orthonormal set of vectors. Therefore,
$\langle o_j, o_k \rangle = 0$ if $j \neq k$, and $\langle o_k, o_k \rangle = 1$. Thus, the only terms of the double
sum which are nonzero are when $j = k$ so we have a single sum,

$$\|f\|^2 = \sum_k a_k^2 = \sum_k |\langle f(t), o_k \rangle|^2$$

which is the first portion of what we were going to prove. The inner product identity can be proven in an identical fashion to that above, so we leave that as an exercise.

2.3.1   The Isometry between $L^2[a, b]$ and l^2

We will now explore the idea of an isometry. Stated plainly, an isometry is a rule, or transform, which maps one Hilbert space into another Hilbert space, while preserving two things: a) all distance and b) all inner products. A simple example in the two-dimensional plane R^2 is a rotation by 90°, or by any fixed number of degrees. Since the plane remains fixed, with all of the vectors rotating the same number of degrees, it is an isometry. Other examples are flips about the x and y axis, or any other line through the origin.

We will now introduce a couple of definitions which will allow us to express ourselves more easily in the future. We begin the formal definition of a linear operator.

Definition 2.3.1 (Linear Operator) *A linear operator, or linear map, from one Hilbert space H to another space G is a rule which uniquely assigns an element $g \in G$ to each $h \in H$. In addition, it must be linear. We will generally denote these by $\mathcal{L} : H \to G$, meaning that $\mathcal{L}(f) = g$, with $f \in H$, and $g \in G$. Moreover, linearity means that $\mathcal{L}(c_1 h_1 + c_2 h_2) = c_1 \mathcal{L}(h_1) + c_2 \mathcal{L}(h_2)$.*

We will now formally define an isometry between Hilbert spaces.

Definition 2.3.2 (Isometry) *Let $\mathcal{L}$ be a linear operator from one Hilbert space H to another Hilbert space G, i.e., $\mathcal{L} : H \to G$. Furthermore, let us denote the inner products in H and G by $\langle, \rangle_H$ and $\langle, \rangle_G$. Then, we say that $\mathcal{L}$ is an isometry if and only if for any $f, g \in H$, we have*

$$\langle f, g \rangle_H = \langle \mathcal{L}(f), \mathcal{L}(g) \rangle_G.$$

The fundamental nature of an isometry is that you can measure distances and angles for functions in one space, say H, by measuring the distances and angles of the images of the functions in the corresponding space, G. This is critical for Fourier Analysis. A great deal of the reason why we use Fourier Series, and in the future the Fourier Transform is that we can often measure something easily in one space, and not so easily in another space. Thus, we choose the place where things are easiest, and then the result extends to the more difficult space. This allows us to analyze the relationships between similar functions, images, or objects in two different ways, and then choose the way which is most clear.

We now define the critical Hilbert space l^2. We are interested in sequences of constants which are either real or complex. We will denote these for now by $\{c_k\}$. For the purposes of this book, we will generally have k be either $k = 0, 1, 2, 3 \ldots \infty$, or $k = -\infty \cdots - 2, -1, 0, 1, 2, \ldots \infty$.

Definition 2.3.3 (Definition of l^2) *Let $\{c_k\}_k$ be a countable sequence of constants, which may be either real or complex. We say that $\{c_k\}_k \in l^2$ if $\sum_k |c_k|^2 < \infty$. Furthermore, we define the inner product on l^2 to be*

$$\langle \{a_k\}, \{b_k\} \rangle = \sum_k a_k, \bar{b}_k.$$

Note then that the squared distance on l^2 is simply given by $\sum_k |a_k|^2$.

The notation in the inner product $\bar{b}$ is the complex conjugate of b. Namely, if b is complex, $b = \alpha + i\beta$, then $\bar{b} = \alpha - i\beta$. This is necessary so that the length of complex sequences is positive. Suppose that a sequence has only one nonzero element b, then its squared length would be $\|b\|_2^2 = b\bar{b} = (\alpha + i\beta)(\alpha - i\beta) = \alpha^2 + \beta^2$.

We now will investigate the critical idea of this section. We define by $\mathcal{F}$ the linear operator which is given by the definition of the Fourier Series. Namely, the orthonormal representation in (2.15) states that

$$f(t) = \frac{a_0}{\sqrt{2\pi}} + \sum_{k=1}^{\infty} a_k \frac{\cos(kt)}{\sqrt{\pi}} + b_k \frac{\sin(kt)}{\pi}, \tag{2.17}$$

where

$$a_k = \int_{-\pi}^{\pi} \frac{1}{\sqrt{\pi}} \cos(kt) f(t) dt, \text{ and } b_k = \int_{-\pi}^{\pi} \frac{1}{\sqrt{\pi}} \sin(kt) f(t) dt,$$

for $k \geq 0$ and

$$a_0 = \int_{-\pi}^{\pi} \frac{1}{\sqrt{2\pi}} f(t) dt.$$

We therefore formally define $\mathcal{F}$ to be the operator mapping a function $f(t) \in L^2[a, b]$ to its orthonormal coefficients $\{a_k\}_{k=0}^{\infty}$, and $\{b_k\}_{k=1}^{\infty}$. While at first appearance this looks like $\mathcal{F}$ is mapping $f(t)$ to two sequences in l^2, we do not look at it that way. They are two joined sets of coordinates which we associate with the cosine and sine terms, respectively. We say that $a_{-k} = b_k$, and then clearly we have exactly one sequence in l^2. We will be switching notation in the future, partially because of the problem between the dual cosine and sine coefficients.

We now state the critical theorem of this section, which is certainly one of the critical ideas of Fourier Analysis.

Theorem 2.3.2 (The Fourier Isometry) *The Fourier Series operator $\mathcal{F}$ defined above is an isometry between $L^2[a, b]$ and l^2. Specifically, let $f(t)$ be an element of $L^2[a, b]$. Then, we have*

$$\|f(t)\|_2^2 = \int_a^b |f(t)|^2 dt = \sum_{k=-\infty}^{\infty} |a_k|^2, \tag{2.18}$$

where we remember that for $k < 0$, $a_k = b_{-k}$.

Proof: It turns out that we have already proven this theorem. Note that the Pythagorean Theorem 2.3.1 guarantees us that anytime you have a sequence of complete orthonormal functions in a Hilbert space, you can calculate the length or distance of a function, and inner products between two functions, from the orthogonal coefficients of the function.

Thus, the Pythagorean theorem states that any set of orthonormal functions naturally generates an isometry between the elements of the original Hilbert space, and the Hilbert space $\{l^2\}$.

2.3.2   Complex Notation

We will now switch into complex notation for Fourier Series. We will stay with this notation for most of the rest of this book. The purpose is to eliminate the need to separate the cosine coefficients $\{a_k\}$ and sine coefficients $\{b_k\}$ and the confusion in how the above described Fourier isometry is notated. Recall that if $i^2 = -1$, then the complex exponential is defined as

$$e^{\alpha+i\beta} = exp(\alpha + i\beta) = exp(\alpha)(\cos(\beta) + i\sin(\beta)).$$

For the purposes of this book, we will generally be interested in

$$exp((\alpha + i\beta)t) = exp(\alpha t)(\cos(\beta t) + i\sin(\beta t).$$

First for the purpose of understanding this, let us let $\alpha = 0$, and take two derivatives of $exp(i\beta t) = e^{i\beta t}$ from using the definition of the derivative of the exponential, and the definition of the derivatives of the cosine and sine terms on the right. Note that $\frac{d^2}{dt^2} exp(i\beta t) = i^2\beta^2 exp(i\beta t) = (-1)\beta^2 exp(i\beta t)$ using the definition of the exponential. In addition, you get $\frac{d^2}{dt^2} exp(i\beta t) = -1\beta^2 \cos(\beta t) - \beta^2 \sin(\beta t)$. Thus, the notation is not artificial, but obeys the rules of differentiation and addition. There are several exercises so if you are new to this you can become familiar.

We will now start using $exp(ikt) \equiv \cos(kt) + i\sin(kt)$ as the standard functions on $[-\pi, \pi]$. Now, we will state the Fourier Series theorem in complex notation.

Theorem 2.3.3 *Let $f(t)$ be any function in $L^2[-\pi, \pi]$. Then, we can represent $f(t)$ in a series as*

$$f(t) = \frac{1}{\sqrt{2\pi}} \sum_{k=-\infty}^{\infty} c_k e^{ikt} \tag{2.19}$$

where

$$c_k = \frac{1}{\sqrt{2\pi}} \langle f(t), e^{ikt} \rangle.$$

Note first of all that this is much simpler than Theorem 2.1.1 (with the exception of getting used to complex notation). There is only one set of coefficients instead of two. There is only one set of functions (rather than cosines and sines). Also note that $exp(ikt) + exp(-ikt) = 2\cos(kt)$ and $exp(ikt) - exp(-ikt) = 2\sin(kt)$, so the cosine and sine coefficients and functions are easily recovered from the new functions and coefficients (which will generally be complex).

We must also redefine the inner product for complex-valued functions at this time.

Definition 2.3.4 (Complex Inner Product) *The inner product for complex functions of a single variable t on $L^2[a, b]$ is given by*

$$\langle f(t), g(t) \rangle = \int_a^b f(t) \overline{g}(t) dt.$$

Note that this corresponds with the definition on l^2.

One thing to keep in mind with the complex inner product is that it is linear in the first argument ($f(t)$ above) but conjugate linear in the second argument ($g(t)$ above), i.e., $\langle f(t), cg(t) \rangle = \overline{c} \langle f(t), g(t) \rangle$. This is necessary to make distances positive for complex vectors or functions.

We now restate the Fourier isometry from $L^2[a, b]$ to l^2 using the complex notation, which we will tend to use from now on.

Definition 2.3.5 (Fourier Operator) *Given any function $f(t) \in L^2[a, b]$ and its complex representation (2.19)*

$$f(t) = \frac{1}{\sqrt{2\pi}} \sum_{k=-\infty}^{\infty} c_k e^{ikt}$$

we define $\mathcal{F} : L^2[a, b] \to l^2$ by

$$\mathcal{F}(f(t)) = \{c_k\}_k$$

where we have noted that $\{c_k\}_k \in l^2$.

Theorem 2.3.4 (Fourier Series Isometry) *The Fourier Series operator* $\mathcal{F}$ *defined above is an isometry between* $L^2[a,b]$ *and* l^2. *Specifically, let* $f(t)$ *and* $g(t)$ *be elements of* $L^2[-\pi, \pi]$. *Moreover, let* $f(t)$ *and* $g(t)$ *have the representation*

$$f(t) = \frac{1}{\sqrt{2\pi}} \sum_{k=-\infty}^{\infty} c_k e^{ikt},$$

and

$$g(t) = \frac{1}{\sqrt{2\pi}} \sum_{k=-\infty}^{\infty} d_k e^{ikt}.$$

Then, we have the lengths being equal in both spaces, i.e., $\|f(t)\|^2 = \sum_k |c_k|^2$, $\|g(t)\|^2 = \sum_k |d_k|^2$, *and in addition that the inner product*

$$\langle f(t), g(t) \rangle = \int_a^b f(t)\overline{g}(t)dt = \sum_k c_k \overline{d}_k,$$

can be calculated in either space. Thus, angles and lengths are preserved between the two spaces.

We can state the above theorem for $L^2[a,b]$, by merely changing the expansions. Thus, this holds on $L^2[a,b]$.

Proof: The first portion is just a restatement of the Pythagorean theorem. The second portion follows from direct substitution.

Differences between Notations

We know from Theorem (2.3.1) that whenever we have a complete orthonormal set of functions, we have a corresponding orthonormal expansion. The question arises, "Why do we have two for Fourier Series, and how does that change?" Both are the same, and we just want you to recognize the difference.

Specifically, from Theorem 2.1.1, we can write

$$f(t) = \frac{a_0}{2} + \sum_{k=1}^{\infty} a_k \cos(kt) + b_k \sin(kt)$$

where

$$a_k = \frac{1}{\pi} \int_{-\pi}^{\pi} f(t) \cos(kt)dt, \text{ and } b_k = \frac{1}{\pi} \int_{-\pi}^{\pi} f(t) \sin(kt)dt.$$

But we later introduced the representation in (2.15)

$$f(t) = \frac{a_0}{\sqrt{2\pi}} + \frac{1}{\sqrt{\pi}} \sum_{k=1}^{\infty} a_k \cos(kt) + b_k \sin(kt),$$

where

$$a_k = \int_{-\pi}^{\pi} \frac{1}{\sqrt{\pi}} \cos(kt) f(t) dt, \text{ and } b_k = \int_{-\pi}^{\pi} \frac{1}{\sqrt{\pi}} \sin(kt) f(t) dt,$$

for $k \geq 1$ and $a_0 = 1/\sqrt{2\pi} \int f(t) dt$. They are the same. Note that in the second representation, the $\sqrt{\pi}$ is split, and in the first, it is combined. Why should we have two notations? The representation in (2.15) is necessary to make sure the isometry is preserved. The original representation in Theorem 2.1.1 is somewhat cleaner. Moreover, you will eventually see different representations and you need to know how they relate. They are all the same, but the notation is changed.

To make this clear, let us consider the following project. Given a complete and orthogonal set of functions $\{f_k\}$ in a Hilbert space $\mathcal{H}$, we want to have orthonormal expansions. This means we must normalize the functions. We do this very simply by defining

$$o_k = \frac{f_k}{\|f_k\|}.$$

Now, for any function $f \in \mathcal{H}$, we have

$$f = \sum_k \langle f, o_k \rangle o_k$$

$$= \sum_k \langle f, \frac{f_k}{\|f_k\|} \rangle \frac{f_k}{\|f_k\|} \tag{2.20}$$

$$= \frac{1}{\|f_k\|^2} \sum_k \langle f, f_k \rangle f_k. \tag{2.21}$$

The difference in representations between (2.21) and (2.20) is exactly the difference between the Fourier references above. To obtain the isometry, we have to use (2.20) and the coefficients of (2.20).

2.3.3 Estimating Truncation Errors

Only under rare circumstances are we able to turn infinite series into a concrete equation. Thus, we are often forced to estimate an infinite Fourier Series with its finite approximation. Let us refer to the Fourier summation as $S(f)$, and the truncated summation $S_N(f)$, and the error in truncation as $E_N(f)$, so we have

$$
\begin{aligned}
f(t) = S(f) &= \frac{a_0}{2} + \sum_{k=1}^{\infty} a_k \cos(kt) + b_k \sin(kt) \\
&= \left(\frac{a_0}{2} + \sum_{k=1}^{N} a_k \cos(kt) + b_k \sin(kt) \right) + \left(\sum_{k=N+1}^{\infty} a_k \cos(kt) + b_k \sin(kt) \right) \\
&= S_N(f) + E_N(f).
\end{aligned}
\tag{2.22}
$$

Similarly if we are using complex notation, let us define

$$
\begin{aligned}
f(t) = S(f) &= \frac{1}{\sqrt{2\pi}} \sum_{k=-\infty}^{\infty} c_k e^{ikt} \\
&= \left(\frac{1}{\sqrt{2\pi}} \sum_{|k| \leq N} c_k e^{ikt} \right) + \left(\frac{1}{\sqrt{2\pi}} \sum_{|k| > N} c_k e^{ikt} \right) \\
&= S_N(f) + E_N(f).
\end{aligned}
\tag{2.23}
$$

We are primarily interested in estimating $f(t) - S_N(f) = E_N(f)$. Generally, the easiest way to estimate this is in the metric,

$$
\| f(t) - S_N(f) \|_2^2 = \int_{-\pi}^{\pi} |f(t) - S_N(f)|^2 dt = \int_{-\pi}^{\pi} |E_N(f)|^2 dt.
\tag{2.24}
$$

Thus, if we want to estimate our error, we can estimate either of the two quantities at the right above. By the Fourier isometry, each of those quantities can be measured at least two ways.

Estimating $\int_{-\pi}^{\pi} |E_N(f)|^2 dt$ by calculating the integral in time requires the numerical estimation integral, which creates some difficulties, but is possible. Estimating the Fourier Transform via the isometry is sometimes much easier. In addition, creative use of the isometry will allow us to measure this in a number of ways. One common way is to realize that

$$
\| E_N(f) \|^2 = \sum_{|k| > N} |c_k|^2.
$$

Since this is an infinite integral, it is not generally easy to calculate in closed form. However, we know that

$$
\| f \|^2 - \sum_{|k| \leq N} |c_k|^2 = \sum_{k > N} |c_k|^2.
$$

Thus, we can calculate the norm of f by integrating in time directly, and then subtract the finite sum of coefficients until the values on the right are sufficiently small. This is one of the advantages of the isometry. You can measure things in different ways, and some are much more efficient than others.

An Example:

As an example, we will return to the first Fourier Series we calculated. Recall that

$$\chi_\pi(t) = \frac{1}{2} + \sum_{k=0}^{\infty} \frac{2\sin(k\pi/2)}{k\pi} \cos(kt),$$

on $[-\pi, \pi]$ where

$$\chi_\pi(t) = \begin{cases} 1 \text{ if } |t| < \pi \\ 0 \text{ if } |t| > \pi \end{cases}, \tag{2.25}$$

Now, we would like to calculate how many coefficients it will take to have the error $\|E_N(f)\| < .01$. We must first switch to the orthonormal representation

$$\chi_\pi(t) = \frac{\sqrt{2\pi}}{2} \frac{1}{\sqrt{2\pi}} + \sum_{k=0}^{\infty} \left(\frac{2\sin(k\pi/2)}{k\sqrt{\pi}} \right) \frac{\cos(kt)}{\sqrt{\pi}},$$

where we have normalized the functions 1 and $\cos(kt)$ and appropriately adjusted the coefficients. Now, by the isometry, we know that $\|\chi(t)\|^2$ is equal to

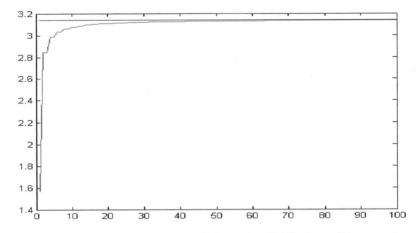

Figure 2.6: We plot the convergence of the series (2.26) above. The error is represented by the difference of the squared integral of the function, at top, and the asymptotically approaching sum of squared coefficients, below. While these two do asymptotically approach each other, this convergence is very slow.

the sum of the squared coefficients, or

$$\int_{-\pi}^{\pi} |\chi_\pi(t)|^2 dt = \pi = \frac{\pi}{2} + \sum_{k=1}^{\infty} \left(\frac{2\sin(k\pi/2)}{k\sqrt{\pi}}\right)^2. \tag{2.26}$$

This leads us to the equality

$$\sum_{k=1}^{\infty} \left(\frac{2\sin(k\pi/2)}{k\sqrt{\pi}}\right)^2 = \sum_{k=1}^{\infty} \frac{4\sin^2(k\pi/2)}{k^2\pi} = \frac{\pi}{2}.$$

The question is, how fast does this happen? By our discussions above

$$\|E_N(f)\|^2 = \pi - \left(\frac{\pi}{2} + \sum_{k=1}^{N} \frac{4\sin^2(k\pi/2)}{k^2\pi}\right). \tag{2.27}$$

This can be quickly determined numerically. In Figure 2.6, we see this above sum converging to π. A quick check shows that the absolute squared error $\|E_N(f)\|^2$ is less than .01 after 64 terms. Recall, however, that $\|E_N(f)\|^2$ is much smaller than $\|E_N(f)\|$. To check $\|E_N(f)\|$, we must take the square root of (2.27), and we find that we need 6368 terms. Thus, this sum converges very slowly, if you want a lot of accuracy. This error level would be very hard to check with a direct numerical integration.

2.3.4 Problems and Exercises:

1. Prove that if $f(t)$ and $g(t) \in L^2[a,b]$, then for any constants $h(t) = \alpha f(t) + \beta g(t) < \infty$.

2. Use basic trigonometric identities to verify that $e^{ix}e^{iy} = e^{i(x+y)}$.

3. Prove the inner product statement in Theorem 2.3.1.

4. Show that the representation in Theorem 2.3.3 is a valid orthonormal representation. Specifically, show that the functions

$$\frac{1}{\sqrt{2\pi}} e^{ikt}$$

are orthonormal on $[-2\pi, 2\pi]$.

5. Find the orthonormal representation for Theorem 2.1.2, such as we rewrote Theorem 2.1.1 in Theorem 2.3.3. First, find the appropriate complex exponentials for $[a,b]$, as are suggested by the sine and cosines of Theorem 2.1.2. Make sure that your altered complex exponentials are orthogonal. Add in the normalization factors which will make them

orthonormal. Finally, write the final form of the new theorem, as in Theorem 2.3.3.

6. Verify that the Fourier isometry holds on $[-\pi, \pi]$ for $f(t) = t$. To do this, (a) calculate the coefficients of the orthogonal Fourier Series from representation in (2.17), (b) calculate the sum of the squared coefficients, and (c) calculate the norm of the function as $\int_{-\pi}^{\pi} |f(t)|^2 dt$. How many terms in the Fourier Series are necessary to have the isometry be under 5%? How many until you are under 3%, or 1%?

7. Verify that the Fourier isometry holds on $[-\pi, \pi]$ for $f(t) = \chi_{\pi/4}(t)$. To do this, (a) calculate the coefficients of the orthogonal Fourier Series from representation in (2.17), (b) calculate the sum of the squared coefficients numerically, or analytically if possible, and (c) calculate the norm of the function as $\int_{-\pi}^{\pi} |f(t)|^2 dt$. How many terms in the Fourier Series are necessary to have the isometry be under 5%? How many until you are under 3%, or 1%?

8. Verify that the Fourier isometry holds on $[-\pi, \pi]$ for $f(t) = t^2$. To do this, (a) calculate the coefficients of the orthogonal Fourier Series from representation in (2.17), (b) calculate the sum of the squared coefficients numerically, or analytically if possible, and (c) calculate the norm of the function as $\int_{-\pi}^{\pi} |f(t)|^2 dt$. How many terms in the Fourier Series are necessary to have the isometry be under 5%? How many until you are under 3%, or 1%?

2.4 Differentiation and Convergence Rates

We now get to a basic idea of Fourier Analysis. It turns out that the Fourier isometry makes this an easy exercise. Let us suppose that we have

$$f(t) = \sum_{k=-\infty}^{\infty} c_k e^{ikt},$$

as above. The question comes up, can we differentiate $f(t)$ in a term by term manner, namely is

$$f'(t) = \sum_{k=-\infty}^{\infty} (ik) c_k e^{ikt}? \tag{2.28}$$

For the purpose of this book, we are only interested in these functions as functions of the space $L^2[-\pi, \pi]$. A simple answer to this question is that if the sum on the right only has a finite number of terms, then f' can be represented as the above series always. The linearity of the derivative, or the fact that $\frac{d}{dt}(af(t) + bg(t)) = af'(t) + bg'(t)$, guarantees this. This becomes trickier when the sum has an infinite number of terms, which is almost always the case.

Let us look at this question as a question about the isometry. Namely, instead of considering $f(t)$ and whether or not it has derivatives in $L^2[a, b]$ let us consider the question, "Do the Fourier coefficients corresponding to $f'(t)$ exist in l^2?" or is

$$\|f'(t)\|^2 = \sum_{k=-\infty}^{\infty} |(ik)^2 c_k|^2 = \sum_{k=-\infty}^{\infty} k^2 |c_k|^2 < \infty. \tag{2.29}$$

If the sum at the right is finite, then there certainly is a function, which we will call h in $L^2[-\pi, \pi]$ which has that Fourier expansion. The remaining question is whether that function is $f'(t)$?

2.4.1   A Quandary between Calculus and Fourier Analysis

Let us remember integration by parts, where $\int f'g + fg'dt = fg$ or $\int f'g = fg - \int fg'dt$. Remember the definition of the complex inner product, and the coefficients c_k, which gives us the coefficients d_k for $f'(t)$ as

$$d_k = \langle f'(t), \overline{e^{ikt}} \rangle = \int f'(t)e^{-ikt}dt$$
$$= f(t)e^{ikt}|_{-\pi}^{\pi} + \int f(t)(ik)e^{-ikt}dt$$
$$= f(t)e^{ikt}|_{-\pi}^{\pi} + (ik)c_k$$
$$= (-f(\pi) + f(-\pi)) + (ik)c_k.$$

Thus, the coefficients of h, which is the derivative of the Fourier Series of f, would differ from those of f' if $f(\pi) \neq f(-\pi)$. This is where the quandary of Fourier Analysis and $L^2[-\pi, \pi]$ becomes a little tricky. Let us define $f_1(t) = f(t)$ everywhere except at π and $-\pi$, where $f_1(t) = 0$ for the sake of argument. Then, the Fourier coefficients for $f_1(t)$ would be identical to those of h. But $f(t)$ and $f_1(t)$ are identical except at two points, thus

$$\|f(t) - f_1(t)\|^2 = \int |f(t) - f_1(t)|^2 dt = 0$$

so they are the same function in $L^2[a, b]$. Thus, one could argue that the coefficients of h are the coefficients of f.

This is not a valid argument for functions in $L^2[-\pi, \pi]$. If we were dealing with only the continuous functions, then the argument would be valid. The problem is that the integration by parts above assumes that $f'(t) \in L^2[-\pi, \pi]$, but nothing more about $f'(t)$. We utilized the Fundamental Theorem of Calculus, but that assumes that $f'(t)$ is continuous. A very extensive resolution of this quandary is given in Tolstov's book [20]. The theorems of Jackson and other approaches are elegantly described in Cheney [5]. The above argument is "morally" correct, and the details have been extensively studied. The reader is encouraged to look more extensively into these details. For the purposes of this book, we will concentrate on the criterion of equation (2.29). This is sufficient to understand Fourier Series, derivatives, and the rates of decay of Fourier Series, which are essential for the applications in later chapters.

2.4.2 Derivatives and Rates of Decay

Let us return to a basic exponential series

$$f(t) = \sum_{k=-\infty}^{\infty} c_k e^{ikt}$$

and differentiate it in a term by term manner to obtain the hopeful representation (2.28)

$$f'(t) = \sum_{k=-\infty}^{\infty} (ik) c_k e^{ikt}.$$

The function $f'(t)$ as represented in (2.28) can only exist in $L^2[-\pi, \pi]$ if

$$\|f'(t)\|^2 = \sum_{k=-\infty}^{\infty} |(ik)^2 c_k|^2 = \sum_{k=-\infty}^{\infty} k^2 |c_k|^2 < \infty.$$

Let us consider the consequences of this equation. Recall the key factors for a sequence to converge. Most notably consider the sum of the series $\sum_k \frac{1}{k} = \infty$. This can be proven by any number of methods. More generally, we can say that

$$\sum_{k=1}^{\infty} \frac{1}{k^{1+\epsilon}}$$

is finite whenever $\epsilon > 0$ and is infinite whenever $\epsilon \leq 0$. Thus, the dividing line between convergence and divergence is $\frac{1}{k}$.

Now, let us consider some notation. We will use this to consider more general sequences, or to compare one sequence to another. This will be necessary to understand the nature of Fourier coefficients.

Definition 2.4.1 *Given a reference sequence $\{b_k\}$, we say that another sequence $\{a_k\}$ is $O(b_k)$ if for some finite M the inequality*

$$\frac{a_k}{b_k} < M < \infty,$$

holds for all k. The above definition is oftentimes replaced by the limit superior (which we will not go into) and is equivalent to the above. Thus, an equivalent definition is

$$\overline{\lim}_k \frac{a_k}{b_k} < M < \infty.$$

Moreover, we say that $\{a_k\}$ is $o(b_k)$ if

$$\lim_{k \to \infty} \frac{a_k}{b_k} = 0.$$

These are terms as little "o" and big "O" notation. Thus, if one sequence is smaller than another in the limit, the little "o" applies implying one is smaller than the other. If the sequences are comparable, then the "O" notation applied, implying that they are of similar magnitude, up to a constant M of some size (which might be huge). Note that $O(b_k)$ is weaker than $o(b_k)$, so anytime something is little "o" to another sequence it is also big "O", with the constant M being zero. The little "o" notation is therefore stronger and preferable when available.

Returning to our equation (2.29), we have that

$$\sum_{k=-\infty}^{\infty} k^2 |c_k|^2 < \infty. \tag{2.30}$$

if the derivative representation $f'(t)$ exists in $L^2[a, b]$. This implies that $k^2|c_k|^2 = o(1/k)$, or that series $\{k^2|c_k|^2\}$ is asymptotically smaller than the series $1/k$ since the sum of the terms $1/k$ is infinite. If $k^2|c_k|^2$ were not $o(1/k)$, then the sum of these coefficients would have to be infinite (this is false). Thus, it follows that $|c_k|^2 = o(1/k^3)$ or that

$$c_k = o\left(\frac{1}{k^{3/2}}\right). \tag{2.31}$$

The previous argument is morally correct, but not true. To see this let us define a sequence c_k to be 0 whenever $k \neq n^2$ for some n where n is an integer. Let $c_k = 1/k = 1/n^2$ whenever $k = n^2$ with n an integer. Then, the condition in (2.30) would be valid, since the terms would be $\sum_n 1/n^2 < \infty$.

There are many different varieties of the statements for defining when a function is differentiable in terms of its coefficients. Equation (2.30) is correct. Trying to move it to limit equations on the c_ks becomes difficult mathematically and has led to an incredible number of true theorems. They are all within a very small ϵ of the general rule, which is illustrated in (2.31). They are all different, however, and cause confusion. Equation (2.30) is a very good and true guideline. One is referred to [5, 20] and many other books and publications for the details of more exact relationships.

Higher Derivatives

Let us now consider the higher derivatives. Returning to our basic complex Fourier Series

$$f(t) = \sum_{k=-\infty}^{\infty} c_k e^{ikt},$$

we can differentiate this n times in a term by term manner to obtain a series for

$$f^{(n)}(t) = \sum_{k=-\infty}^{\infty} (ik)^n c_k e^{ikt}.$$

This function does not exist in $L^2[-\pi, \pi]$, however, unless

$$\sum_{k=-\infty}^{\infty} k^{2n} |c_k|^2 < \infty.$$

This statement is correct and flows both ways. We have a theorem, specifically

Theorem 2.4.1 *Let $f(t) \in L^2[-\pi, \pi]$ have the Fourier representation*

$$f(t) = \sum_{k=-\infty}^{\infty} c_k e^{ikt}.$$

Then, the nth derivative of $f(t)$, $f^{(n)}(t)$ exists in $L^2[-\pi, \pi]$ if and only if

$$\sum_{k=-\infty}^{\infty} k^{2n} |c_k|^2 < \infty \tag{2.32}$$

and then it has the representation

$$f^{(n)}(t) = \sum_{k=-\infty}^{\infty} (ik)^n c_k e^{ikt}.$$

Let us investigate the consequences of the series in (2.32). As before, a good rule of thumb is that the series $k^{2n}|c_k|^2$ should be asymptotically smaller than the series $1/k$, since the sum of $1/k$ is infinite. This would imply that $k^{2n}|c_k|^2 = o(1/k)$, or that $|c_k|^2 = o(1/k^{2n+1})$, implying that $|c_k| = o(1/k^{n+1/2})$. While this is essentially correct, one can make up counterexamples which as we did in before showing that we can have convergence without this condition.

We can state something more definitive, without being exhaustive on all of the conditions necessary to have a term by term derivative.

Theorem 2.4.2 *Let $f(t) \in L^2[-\pi, \pi]$ have the Fourier representation*

$$f(t) = \sum_{k=-\infty}^{\infty} c_k e^{ikt}.$$

Then, the nth derivative of $f(t)$, $f^{(n)}(t)$ exists in $L^2[-\pi, \pi]$ if

$$|c_k|^2 = o(1/k^{2n+1+\epsilon}) \tag{2.33}$$

for any $\epsilon > 0$, and it will have the valid Fourier representation

$$f^{(n)}(t) = \sum_{k=-\infty}^{\infty} (ik)^n c_k e^{ikt}.$$

This clears up some of the problems. We do not claim to be exhaustive on all of the possible conditions which may make term by term differentiation work. Once again, please refer to the references at the end of this book, and the many other possible Fourier Analysis resources. We believe that this basic understanding is enough to move forward.

2.4.3 Fourier Derivatives and Induced Discontinuities

We will now investigate the decay rates via some examples. We first recall that via our discussion in Section 2.4.1, a function $f(t) \in L^2[-\pi, \pi]$ will not have a valid term by term Fourier representation of its derivative unless $f(\pi) = f(-\pi)$. We would also like to have the Fundamental Theorem of Calculus hold, or we would like

$$f(t) = f(-\pi) + \int_{-\pi}^{t} f'(x)dx.$$

Using this as our definition of a Fourier derivative, we will now analyze several examples to see if term by term differentiation is consistent with this definition, and the decay rates are appropriate.

First Example: Let us consider the simple example of $f(t) = t$ on $[-\pi, \pi]$. While this function seems to be absolutely continuous, its periodic extension, illustrated in Figure 2.4, is discontinuous at every odd multiple of π. It fails the basic test $f(\pi) = f(-\pi)$?, and thus the decay rate cannot be fast. We will leave the prediction of this decay rate and the verification as an exercise.

Second Example: We will now consider the second simple example of $f(t) = t^2$ on $[-\pi, \pi]$. This function does satisfy $f(\pi) = f(-\pi)$ and therefore

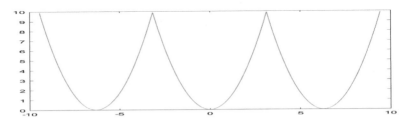

Figure 2.7: We have plotted the periodic extension of $f(t) = t^2$ from $[-\pi, \pi]$ to $[-3\pi, 3\pi]$. Note that the periodic extension is continuous.

since it is continuous, and its derivative is $f'(t) = 2t$, we expect its series to converge much faster. Certainly, this satisfies the Fundamental Theorem of Calculus. Once again, we will let the reader predict the decay rate, and verify this with the Fourier expansion.

2.4.4 Problems and Exercises:

1. (a) Predict the decay rate of the coefficients for the Fourier expansion of $f(t) = t$ on $[-\pi, \pi]$. (b) Calculate the Fourier Series for this and compare this to your prediction. (c) Were you correct?

2. (a) Predict the decay rate of the coefficients for the Fourier expansion of $f(t) = t^2$ on $[-\pi, \pi]$. (b) Calculate the Fourier Series for this and compare this to your prediction. (c) Were you correct?

3. Consider the function

$$f(t) = \begin{cases} t+1 & \text{for } t \in [-1, 0] \\ 1-t & \text{for } t \in [0, 1] \end{cases} \tag{2.34}$$

(a) Does $f(t)$ have a valid Fourier derivative in $[-1, 1]$? (b) What is that derivative if it exists? (c) Calculate the Fourier Transform of the function $f(t)$ on $[-1, 1]$. (d) Take the term by term Fourier Transform of this function. (e) Plot this function and state whether it is consistent with the derivative of the function.

4. **Challenging:** Consider the function $g(t) = f(t)^2$ where

$$f(t) = \begin{cases} t+1 & \text{for } t \in [-1,0] \\ 1-t & \text{for } t \in [0,1] \end{cases} \qquad (2.35)$$

(a) Does $g(t)$ have a valid Fourier derivative in $[-1,1]$? (b) What is that derivative if it exists? (c) Calculate the Fourier Transform of the function $g(t)$ on $[-1,1]$. (d) Take the term by term Fourier Transform of this function. (e) Plot this function and state whether it is consistent with the derivative of the function.

5. Consider the function

$$f(t) = \begin{cases} 0 & \text{for } |t| \geq \pi/2 \\ \cos(t) & \text{for } |t| \leq \pi/2 \end{cases} \qquad (2.36)$$

(a) Does $f(t)$ have a valid Fourier derivative in $[-\pi, \pi]$? (b) What is that derivative if it exists? (c) Calculate the Fourier Transform of the function $f(t)$ on $[-\pi, \pi]$. (d) Take the term by term Fourier Transform of this function. (e) Plot this function and state whether it is consistent with the derivative of the function.

2.5 Sine and Cosine Series

We have talked about Fourier Series on $L^2[a, b]$. By that we are implicitly using both sines and cosines. We have talked about adjusting the sines and cosines to the length of the interval. In addition, we have also talked about the fact that an even function, when expanded about 0, will have only cosine terms, and an odd function, expanded about 0, will have only sine terms.

We will now look to expand functions which are neither odd, nor even, in either cosine or sine series. This cannot be done in the manner described until now in this chapter.

To do this, let us return to an example we had earlier, namely $f(t) = t$. We expanded this on $[-\pi, \pi]$ using the Fourier Series in Figure 2.4. Since this function was odd, this expansion involved only sine terms. Now, we would like to consider expanding it on the half interval $[0, \pi]$. The earlier sine expansion will obviously still converge on this interval.

We can also expand this function on $[0, \pi]$ using cosine terms, should we choose. We do this by considering the function $g(t) = |t|$ on $[-\pi, \pi]$ and expanding it in a traditional Fourier Series. Because $g(t)$ is even, this expansion will have only cosine terms. But since $g(t) = f(t)$ on $[0, \pi]$, this cosine expansion will converge to $f(t)$ on $[0, \pi]$.

Thus, we have a way to expand a function $f(t)$ on $[0, \pi]$ using either a cosine or sine series.

- If you want to expand in a cosine series, you define

$$g_e(t) = \begin{cases} f(t) & \text{if } |t| > 0 \\ f(-t) & \text{if } |t| < 0 \end{cases}, \qquad (2.37)$$

Thus, you are artificially creating a $g(t)$ which is an even extension of $f(t)$.

- If you want to expand in a sine series, you define

$$g_o(t) = \begin{cases} f(t) & \text{if } |t| > 0 \\ -f(t) & \text{if } |t| < 0 \end{cases}, \qquad (2.38)$$

Thus, you are artificially creating a $g(t)$ which is an odd extension of $f(t)$.

The Fourier Series for both of these functions will converge on $[-\pi, \pi]$, and thus will both be equal to $f(t)$ on $[0, \pi]$ since $g_e(t) = g_o(t) = f(t)$ for $t \in [0, \pi]$.

The Fourier coefficients will merely be twice the coefficients they would be in the original series, or we have a cosine series which is

$$f(t) = \frac{a_0}{2} + \sum_{k=1}^{\infty} a_k \cos(kt)$$

and

$$f(t) = \sum_{k=1}^{\infty} b_k \sin(kt),$$

where

$$a_k = \frac{2}{\pi} \int_0^{\pi} f(t) \cos(kt) dt, \text{ and } b_k = \frac{2}{\pi} \int_0^{\pi} f(t) \sin(kt)) dt.$$

Example

We will now illustrate simply expanding $f(t) = t$ in a cosine series on $[0, \pi]$. This means that we are implicitly expanding $g(t) = |t|$ on $[-\pi, \pi]$. Our coefficients are calculated simply from the above formula using integration by parts. Thus when $k = 0$

$$a_0 = \frac{2}{\pi} \int_0^{\pi} t \, dt = \frac{2}{\pi} t^2 / 2 |_0^{\pi} = \pi,$$

so the average value is $a_0/2 = \pi/2$ as expected.

$$
\begin{aligned}
a_k &= \frac{2}{\pi} \int_0^\pi t \cos(kt) dt \\
&= \frac{2}{\pi} \left(t \frac{\sin(kt)}{k} \Big|_0^\pi - \int_0^\pi 1 \frac{\sin(kt)}{k} dt \right) \\
&= \frac{2}{\pi} \left(0 - \frac{-\cos(kt)}{k^2} \Big|_0^\pi \right) \\
&= \frac{2}{k^2\pi}(\cos(k\pi) - \cos(0)) = \frac{2}{k^2\pi}(\cos(k\pi) - 1).
\end{aligned}
\tag{2.39}
$$

Note that the terms at the right will be zero for k even. Thus, we have the cosine series

$$
t = \frac{\pi}{2} + \sum_{k=1}^{\infty} \frac{2}{k^2\pi}(\cos(k\pi) - 1)\cos(kt),
\tag{2.40}
$$

which is valid on $[0, \pi]$.

Now, let us examine the partial sum approximations which are shown in Figure 2.8. These are remarkably good, for very few terms. Recalling the approximations to $f(t) = t$ with the sine approximations on $[-\pi, \pi]$, these converge much more quickly. The question is why? The answer is simple. With the sine approximation on $[-\pi, \pi]$, there is an induced jump discontinuity at the endpoint. By periodizing t, you are eliminating the discontinuity, and trading

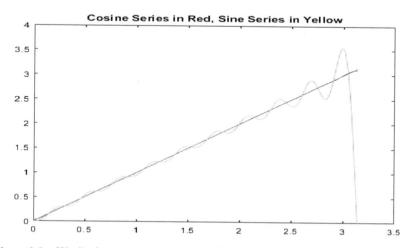

Figure 2.8: We display approximations to $f(t) = t$ with a cosine series and sine series on $[0, \pi]$. Note that with just only 10 terms, the cosine approximation is very good. With sine series is not good with 50 or more terms.

it for only a discontinuity in the derivative. Thus, we are now dealing with a continuous and not a discontinuous function. This makes the approximations much better.

Let us remember at this point that we are approximating an even extension of t on $[0, \pi]$, or the two π periodic extension of t. The odd extension of t would have simply been t, but which is 2π periodic. Thus, let us examine the two periodic extensions, $g_e(t)$ and $g_o(t)$ in Figure 2.9.

Let us go back and examine the cosine representation for t shown in (2.40). An obvious reason that it converges quickly is that the coefficients are $2/k^2$. The sine representation for t (exercise 3 in Section 2.1.3) has coefficients that only decay like $1/k$. Thus, it is natural that this series converges faster. We have talked about series converging at a faster rate if and only if they have derivatives. Let us examine the term by term derivative of equation (2.40) or

$$\frac{d}{dt}\left(\frac{\pi}{2} + \frac{2}{k^2\pi}\sum_{k=1}^{\infty}(\cos(k\pi) - 1)\cos(kt)\right)$$

$$= \frac{2}{k^2\pi}\sum_{k=1}^{\infty}(\cos(k\pi) - 1)(-k\sin(kt))$$

$$= \sum_{k=1}^{\infty}\frac{-2(\cos(k\pi) - 1)}{k\pi}\sin(kt). \tag{2.41}$$

Since the terms are $O(1/k)$, this series should converge, since the sum of its coefficients is finite. In Figure 2.10, we show the first 100 terms of the derivative

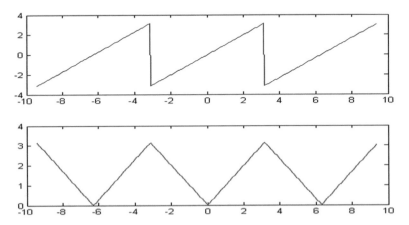

Figure 2.9: We illustrate the even and odd 2π periodic extensions of $f(t) = t$ on $[0, \pi]$ above. Note that while the odd extension $g_o(t)$ might seem more natural, the even extension $g_e(t)$ is continuous while the odd extension has a jump discontinuity. As a result, the cosine series converges much faster on $[0, \pi]$.

of this series. Note that it quite apparently converges for a function which is ± 1. This is exactly the derivative of the extension $g_e(t)$ shown in Figure 2.9.

A key difference between the two extensions in Figure 2.9 is that while they both have discontinuities in their derivatives at isolated points, the odd extension has a jump discontinuity in the function. Note that the even extension is $g_e(t) = |t|$ on $[-\pi, \pi]$ and is periodic after that. If we define

$$h(t) = \begin{cases} 1 \text{ if } 2m\pi < t < (2m+1)\pi \\ -1 \text{ if } (2m+1)\pi < t < 2m\pi \end{cases}, \tag{2.42}$$

then $h(t)$ is the derivative of $g_e(t)$ except at the points $k\pi$. More importantly, we can write that

$$g_e(t) = \int_0^t h(x)dx, \tag{2.43}$$

and this will be valid for all t, even though the derivative is not defined at $m\pi$.

Note that we can similarly define the derivative of the odd extension $g_o(t)$. You will not be able to write an analogue of the equation (2.43), however, due to the jump discontinuity in the function $g_o(t)$.

We summarize by stating the following

Theorem 2.5.1 (Sine and Cosine Representations) *Let $f(t)$ be any function on $L^2[0, T]$. Then, we can represent $f(t)$ in a series on $[0, T]$ as either*

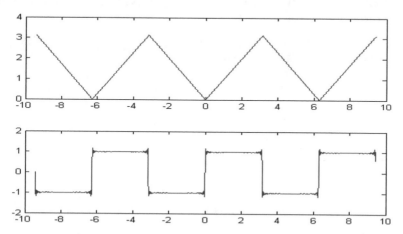

Figure 2.10: We illustrate the first 100 terms of the series for $g_e(t)$ and the derivative of that series above 2.41. First note that the series for $g_e(t)$ converges in the first 10 terms or so, and thus, the 100-term approximation is essentially identical to $g_e(t)$. Second, notice that the derivative is either approximately 1, or -1, according to the derivative of $g_e(t)$. In addition, notice the strong Gibbs ringing at the jump discontinuity, and the slow convergence.

$$f(t) = \frac{a_0}{2} + \sum_{k=1}^{\infty} a_k \cos\left(\frac{k\pi t}{T}\right),$$

or

$$f(t) = \sum_{k=1}^{\infty} b_k \sin\left(\frac{k\pi t}{T}\right)$$

where

$$a_k = \frac{2}{T} \int_0^T f(t) \cos\left(\frac{k\pi t}{T}\right) dt, \quad and \quad b_k = \frac{2}{T} \int_0^T f(t) \sin\left(\frac{k\pi t}{T}\right) dt.$$

Proof: The proof follows directly from the above discussions and the original Fourier Series theorems.

2.5.1 Problems and Exercises:

1. Calculate the sine and cosine series for $f(t) = t$ on $[0, 1]$. Plot the first 30 terms of these series on $[-3, 3]$. Estimate the error between both series and the function on $[0, 1]$ after 30 terms. How many more terms of the sine series are necessary to achieve the same error as was achieved with the cosine series and 30 terms?

2. Plot the first 30 terms of the derivatives of both the sine and cosine series in the above problem. What do you observe? Do either of them converge, and why?

3. Prove that both the cosines $\{1, \cos(kt)\}_{k=1}^{\infty}$, and the sines $\{\sin(kt)\}_{k=1}^{\infty}$ are orthogonal on $[0, \pi]$. Given that

$$\{1, \cos(kt), \sin(kt)\}_{k=1}^{\infty}$$

is complete in $L^2[-\pi, \pi]$ show that both $\{1, \cos(kt)\}_{k=1}^{\infty}$ and $\{\sin(kt)\}_{k=1}^{\infty}$ are complete in $L^2[0, \pi]$.

4. Compute the sine and cosine series for

$$f(t) = \begin{cases} t & \text{if } t \le \pi/2 \\ \pi/2 - t & \text{if } t > \pi/2 \end{cases}, \tag{2.44}$$

on $[0, \pi]$. Which converges faster? Plot the terms. Why do you think this is?

2.6   Perhaps Cosine Series Only

In the last section, we showed how a cosine series, which is naturally even, could far more efficiently approximate an odd function. We would like to push this idea one step further, and try to use only cosine series, which are naturally even, to represent an arbitrary function, which is neither odd nor even.

To begin with, remember that we have already constructed a decomposition of an arbitrary function into its even and odd components. Namely, let us refer to our first Fourier Series formula

$$f(t) = \frac{a_0}{2} + \sum_{k=1}^{\infty} a_k \cos(kt) + b_k \sin(kt).$$

The constant and cosine terms are the even portion of the function, and the sine terms are the odd portion of the function. Thus, we know we can write $f(t) = f_e(t) + f_o(t)$.

Now, we would like to consider writing $f(t) = f_e(t) + f_o(t)$ in terms of odd and even functions, more directly. Since $f_e(t) = f_e(-t)$ and $f_o(t) = -f(-t)$, we have the relations

$$f_e(t) = \frac{1}{2}(f(t) + f(-t))$$

and

$$f_o(t) = \frac{1}{2}(f(t) - f(-t)).$$

Obviously, $f(t) = f_e(t) + f_o(t)$ from the definitions above.

What we showed above in the last section is that an even extension of an odd function is more easily representable than the original odd function. Secondly, we have shown that t^2 is more easily representable than t, or that even functions are more easily representable.

The question then is, can't we make everything even? The answer is yes. We will call this they Compression Series. The algorithm is simple.

Forward Transform:

1. Calculate the cosine series coefficients for $f_e(t)$.

2. Calculate the cosine series coefficients for $f_o(t)$.

3. Figure out how many terms are necessary to represent $f_e(t)$ and $f_o(t)$ within a desired accuracy, and keep only these terms.

The reconstruction algorithm is similarly simple.

Inverse Transform:

1. Calculate $f_e(t)$ within the desired accuracy from the above stored coefficients.

2. Calculate $f_o(t)$ within the desired accuracy on $[0, \pi]$ from the above stored coefficients. Let $f_o(-t) = -f_o(t)$.

3. Calculate $f(t) = f_e(t) + f_o(t)$ within the desired accuracy with very few coefficients.

Theoretically, we have suggested that you can store, or transmit, many fewer coefficients using the above algorithm than blindly using the Fourier Transform. Let us calculate a test example, just to understand what the savings might be. We will use the very simple function $f(t) = 2 + 3t$, on the interval $[-\pi, \pi]$.

The even and odd components of this function are obvious, namely $f_3(t) = 2$ and $f_o(t) = 3t$. Now, let us try to represent this function with a minimal number of coefficients, using both the standard Fourier Transform and the cosine series.

The sine series for $f(t) = t$ on $[-\pi, \pi]$ is

$$t = \sum_{k=1}^{\infty} -\frac{2\cos(k\pi)}{k}\sin(kt). \tag{2.45}$$

By linearity, the Fourier Series for our function

$$f(t) = 2 + 3t = 2 - 3\sum_{k=1}^{\infty}\frac{2\cos(k\pi)}{k}\sin(kt).$$

Let us simplify the discussion by realizing that the only question is "How quickly can be represented the odd portion of the function?"

The cosine series for $f(t) = t$ is

$$t = \frac{\pi}{2} + \sum_{k=1}^{\infty}\frac{2}{k^2\pi}(\cos(k\pi) - 1)\cos(kt), \tag{2.46}$$

One of the series decays as $O(1/k)$ and the other as $O(1/k^2)$. Thus, the cosine series is more efficient. We illustrate while emphasizing that both obey the isometry rule, or Pythagorean theorem. To do this, we must first rewrite these series in their orthonormal form. This isolates the orthonormal coefficients, and the orthonormal functions, or vectors. The orthonormal forms are

$$t = \sum_{k=1}^{\infty}\left(-\frac{2\cos(k\pi)\sqrt{\pi}}{k}\right)\frac{\sin(kt)}{\sqrt{\pi}}. \tag{2.47}$$

and

$$t = \frac{\pi^{3/2}}{\sqrt{2}} \frac{1}{\sqrt{2\pi}} + \sum_{k=1}^{\infty} \left(\frac{2}{k^2 \pi} (\cos(k\pi) - 1) \right) \frac{\cos(kt)}{\sqrt{\pi}}. \qquad (2.48)$$

The obvious question is "Why do we have to rewrite the equations 2.45 and 2.46 in the more cumbersome forms 2.47 and 2.48?" The answer is that we have to have the orthonormal coefficients. Combining constants in the series makes it more pleasant, but does not allow us to check on the Pythagorean identity.

Let us now examine this and the convergence rates of both series. The norm of the function $f(t) = t$ is

$$\int_{-\pi}^{\pi} t^2 dt = \frac{t^3}{3} |_{-\pi}^{\pi} = \frac{2\pi^3}{3} \approx 20.67.$$

The Fourier isometry tells us that we also have the strange set of equalities

$$\frac{2\pi^3}{3} = \frac{\pi^3}{2} + \sum_{k=1}^{\infty} \left(\frac{2}{k^2 \pi} (\cos(k\pi) - 1) \right)^2 = \sum_{k=1}^{\infty} \left(-\frac{2\cos(k\pi)\sqrt{\pi}}{k} \right)^2. \qquad (2.49)$$

Let us check this numerically. In Figure 2.11, you see the sum of the squared coefficients of the cosine series and the sine series. It is apparent from 2.11 that the cosine series converges much quicker. The absolute numbers can be objectively judged by the relative mean squared error RMSE, which is defined to be

$$RMSE = \frac{\|f(t) - S_N(f(t))\|^2}{\|f(t)\|^2}.$$

After 50 terms, the RMSE for the cosine series is 3e-7. The corresponding RMSE for the sine series is .12. After 1000 terms, the sine series has a RMSE of 6e-4. The cosine series achieves this with eight terms. Thus, 1000 coefficients of the sine series are equivalent to eight terms from the cosine series. The savings in computation and storage is very, very significant.

2.6.1 Induced Discontinuities vs. True Discontinuities

The above conversation on using the cosine series focuses on the fact that a Fourier Series is naturally periodic. If the function to be represented is odd, there is an induced jump discontinuity at the endpoint. We have put forward the idea that by changing the odd portion of the function to be even, and then using a cosine representation on it, this can be avoided.

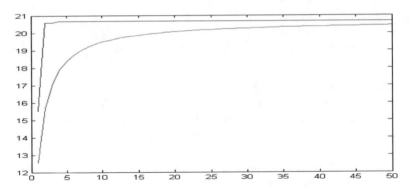

Figure 2.11: The blue curve above is the sum of the squared coefficients from the cosine series 2.48. The green curve is similarly the sum of the squared coefficients from the sine series 2.47. Note that the cosine series converges much faster.

The question which was not addressed is "What should be done if the function has an actual discontinuity in it?" The answer depends upon the application. If you are using the Fourier Series for computation in a differential equation, you probably do not want to erase the discontinuity, since it is fundamental to the answer. On the other hand, if you are just trying to find efficient storage methods, represent jump discontinuities as exactly what they are. Record the jump discontinuities, and then use the Fourier Transform to represent the smoothed function. This is one of many solutions. A simple example is illustrated in Figure 2.12. The function above is

$$f(t) = \begin{cases} -t + 1 \text{ if } |t| < 0 \\ t^2 \text{ if } |t| > 0 \end{cases}. \tag{2.50}$$

Removing the discontinuity at the origin will result in a much more efficient Fourier Series.

2.6.2 Problems and Exercises:

1. **Challenging:** Calculate the Fourier Series for the above function. How many terms are necessary to represent it with a RMSE of less than .01?

2. **Challenging:** Remove the discontinuity from the above function, and represent it using only cosine series, and the removed discontinuity. How many terms are necessary to represent it with a RMSE of less than .01?

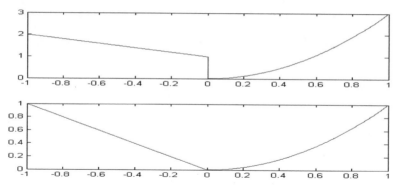

Figure 2.12:   We illustrate a jump discontinuity in the first of the above graphs, and the graph with the jump discontinuity removed below. The standard Fourier Series for this function would converge very slowly with the extension "efficient representation".

2.7   Gibbs Ringing Phenomenon

Throughout all of our Fourier Series examples, we have noticed a ringing phenomenon, at the discontinuities of the functions. This phenomena is referred to as Gibbs ringing in honor of an American mathematician and physicist who wrote about it in the late 1800s [9]. We will address this phenomena from a more analytical perspective in Chapter 4.

We will look into this more thoroughly from the numerical perspective at this time. We will utilize the Fourier Series of the function $\chi_{\pi/2}(t)$ which we showed earlier to have the expansion

$$\chi(t) = \frac{1}{2} + \sum_{k=0}^{\infty} \frac{2\sin(k\pi/2)}{k\pi} \cos(kt).$$

In Figure 2.13, we examine the maximas and minimas of the approximations to this function. We notice that they are all nearly as far above or below the desired function, regardless of the number of terms in the approximation. This is very disturbing, since we have proven that the Fourier Series must converge for all functions in $L^2[a, b]$. It seems that no matter how many terms are added, there is still a large difference between the maxima, minima, and the desired values of 1 and 0.

There are a number of observations that can be made from Figure 2.13. Putting all of them together allows us to understand what seems like an anomaly which is not consistent with our theorems. Let us note some observations.

1. The maxima and minima of the partial sums do not seem to decrease and approach the final desired value.

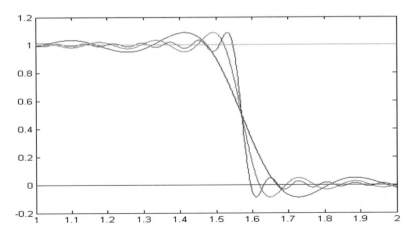

Figure 2.13: We have plotted an approximation to $\chi(t)$ on $[-\pi, \pi]$ above, with 20, 40, and 80 terms. We display this approximation on the interval $[1, 2]$ to get a close-up of the approximations. We notice that the maximum value or peak value of each of the approximations is approximately .1 or so above the desired value of 1. Similarly, the minimum value stays below 0 by a similar amount.

  2. The locations of those maxima and minima seem to get closer to the discontinuity as the number of terms increases.

  3. The area of the error created by this "ringing" seems to decrease.

Let us move from pure speculation based on the graphs, to observation of the numerical facts. The values of the maxima were 1.0899, 1.0896, and 1.0894. The maxima on the right of the discontinuity are located at 1.4140, 1.4920, and 1.5320. Looking at the graphs, it would seem that the maxima will stay at a value around 1.09 and that the location of the maxima will approach $\pi/2 \approx 1.5708$. Let us test this with 200 terms, illustrated in Figure 2.14. The maximum with 200 terms is located at 1.5550, and its value is 1.0895. This would seem to support our ideas put forward above.

We will prove that the above observations demonstrate the truth in Chapter 4. We need more tools to accurately describe this.

2.7.1 Problems and Exercises

- **Gibbs ringing** Calculate the Fourier Series for $\chi_{1/2}(t)$ on $[-1, 1]$, and plot the first 50 terms. Find the maximum value of the series with 30, 40, 50, and 100 terms. Guess at the limit of this maximum value. Find the location of the maximum value on the right of 0, with 30, 40, 50, and 100 terms. Where is the maximum value headed? Finally, estimate the squared error $\int_{-1}^{1} |\chi_{1/2}(t) - S_n(\chi_{1/2})(t)|^2 dt$ and the series on $[-1, 1]$ with the same number of terms (do this numerically).

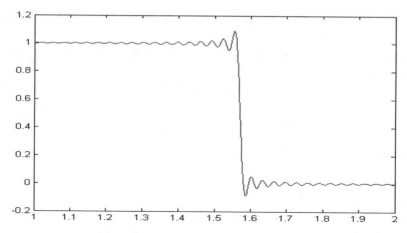

Figure 2.14: We have plotted an approximation to $\chi(t)$ on $[-\pi, \pi]$ above with 200 terms. We display this approximation on the interval $[1, 2]$ to get a close-up of the approximations. Note that the peak is closer to $\pi/2$ than those in Figure 2.13, where there were fewer terms in the sum.

2.8 Convolution and Correlation

There are two very common operations which come up in association with Fourier Series. They are generally associated with localized averages or operations on a function. We will begin by noting that if $f(t) \in L^2[-T, T]$, then the extension of f given by $f = 0$ for $t > T$ allows us to consider $f(t) \in L2[-B, B]$ for any $B > T$. Oftentimes in this section, we will want $B = 2T$. Now, let us define convolution.

Definition 2.8.1 (Convolution and Correlation) *Let $f, g \in L^2[-T, T]$, and by extension $L^2[-2 * T, 2 * T]$. We define the convolution of f and g to be*

$$f * g(t) = \frac{1}{\sqrt{2T}} \int_{-2T}^{2T} f(x)g(t - x)dx. \qquad (2.51)$$

Similarly, we define the correlation of f and g to be

$$f \star g(t) = \frac{1}{\sqrt{2T}} \int_{-2T}^{2T} f(x)g(x - t)dx. \qquad (2.52)$$

We would like to characterize the Fourier Series of these two functions. This leads to the following theorem

Theorem 2.8.1 (Convolution and Correlation Theorem) *Let $f, g \in L^2[-T, T]$. The convolution $f * g$ will be in $L^2[-2T, 2T]$. Furthermore, the Fourier Series of $f * g$ on $L^2[-T, T]$, which is denoted by $\mathcal{F}(f * g)$, is expressed simply by*

$$\mathcal{F}(f * g) = \mathcal{F}(f)\mathcal{F}(g). \tag{2.53}$$

Furthermore, the Fourier Series of the correlation $f \star g \in L^2[-2T, 2T]$ is given by

$$\mathcal{F}(f \star g) = \mathcal{F}(f)\overline{\mathcal{F}(g)}. \tag{2.54}$$

Recall that $\mathcal{F}(f)$ maps the function f to its Fourier coefficients. Thus, the Fourier coefficients of the convolution $f * g$ can be found by pointwise multiplication of the Fourier coefficients of f and g

Proof: We must first demonstrate that $f * g \in L^2[-2T, 2T]$. For simplicity's sake, we will assume that $T = \pi$, since the general proof follows immediately.

The first step is to realize that this is a series of inner products, and they are therefore bounded by the Cauchy–Schwartz inequality. In other words

$$f * g(t) = \int_{-2T}^{2T} f(x)g(t-x)dx = \langle f(x)g(t-x)\rangle \le \|f(x)\|_2 \|g(t-x)\|_2.$$

Another thing to remember is that $f(x) = 0$ for $|x| > T$ and $g(t-x) = 0$ for $|t-x| > T$. Thus, the integral above is zero for $|x| > T$. In addition, if $|t| > 2T$, it would follow that $|t-x| > \pi$ so we have that $f * g(t) = 0$ for $|t| > 2\pi$. Thus, we have that $f * g$ is bounded by the Cauchy–Schwartz theorem, and zero outside of $[-2\pi, 2\pi]$, so the convolution must be square integrable on $[-2\pi, 2\pi]$, and therefore, $f * g \in L^2[-2\pi, 2\pi]$.

We will now prove the second assertion of the theorem. We will do this by simply calculating the Fourier coefficients of the convolution.

The Fourier coefficients are given by

$$
\begin{aligned}
c_k &= \int_{-2\pi}^{2\pi} f * g(t)e^{i(k/2)t}dt = \int_{-2\pi}^{2\pi}\int_{-\infty}^{\infty} f(x)g(t-x)dx e^{i(k/2)t}dt \\
&= \int_{-2\pi}^{2\pi}\int_{-\pi}^{\pi} f(x)g(t-x)dx e^{i(k/2)t}dt \\
&= \int_{-\pi}^{\pi}\int_{-2\pi}^{2\pi} f(x)g(t-x)e^{i(k/2)t}dtdx \\
&= \int_{-\pi}^{\pi} f(x)\int_{-2\pi}^{2\pi} g(t-x)e^{i(k/2)(t-x)}dt e^{(k/2)x}dx. \tag{2.55}
\end{aligned}
$$

Now, if we let $u = t - x$, realize that $g(t-x)$ is only nonzero for $|u| < \pi$ so we get

$$c_k = \int_{-\pi}^{\pi} f(x) \int_{\pi}^{\pi} g(u) e^{i(k/2)u} du e^{(k/2)x} dx$$

$$= \int_{-\pi}^{\pi} f(x) e^{(k/2)x} dx \int_{\pi}^{\pi} g(u) e^{i(k/2)u} du$$

$$= \int_{-2\pi}^{2\pi} f(x) e^{(k/2)x} dx \int_{2\pi}^{2\pi} g(x) e^{i(k/2)x} dx$$

$$= \hat{f}(k)\hat{g}(k), \tag{2.56}$$

where $\hat{f}(k)$ and $\hat{g}(k)$ are the Fourier coefficients on $[-2\pi, 2\pi]$ for f and g.

2.8.1   A couple of classic examples

One of the problems with Fourier Series is that they are very hard to compute. You have to be able to solve the integral equations, which is not easy for most functions. For this reason, we oftentimes use other means to try to calculate the Fourier Series. The convolution and correlation theorems do provide us with one such method.

Example 1: We begin by picking an example, which we can calculate directly, through the integral equations. We are also able to calculate this example by using the convolution equation. This is a very simple example, but gives a very direct idea of what convolution is.

We start with one of our favorite functions $f(t) = \chi_\pi(t)$, on $[-\pi, \pi]$. We want to consider its extension to the whole real line, namely we want to consider it to be zero outside of $[-\pi, \pi]$. Now, we want to consider the convolution $f * f$. Note that in this case, since $\chi(t)$ is an even function, convolution and correlation are equal. We will present the result and then leave the calculations as very important exercises.

The convolution $f * f$ is given by

$$f * f(t) = \begin{cases} 2\pi - |t| & \text{for } |t| < 2\pi \\ 0 & \text{for } |t| > 2\pi \end{cases}. \tag{2.57}$$

This is sometimes called a "hat" function and is illustrated in Figure 2.15. An illustrative picture of the "moving average" or convolution process by which the "hat" function was formed is shown in Figure 2.16.

Example 2: We now consider another example. We choose a simple Gaussian which has been corrupted by noise. We then use a window to do a simple moving average of the function to greatly reduce the noise. This is illustrated in Figure 2.17.

2.8.2 Problems and Exercises:

1. **Gibbs Ringing Revisited** Return to the Fourier Series $\chi_{\pi/2}(t)$, on $[-\pi, \pi]$ which we calculated in 2.8. Calculate a new function, with the Fourier coefficients from 2.8, multiplied by $exp(-k^2/100)$ for each k. Use 30, 50, and 100 terms. Compare the series with and without the added exponential term. Do you see a reduction in Gibbs ringing? Why?

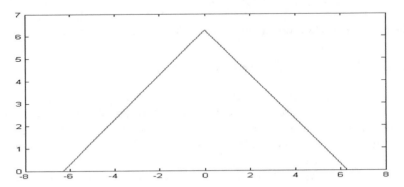

Figure 2.15: The convolution of characteristic functions is illustrated in the first graph above.

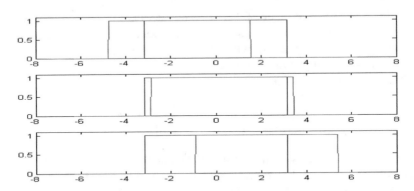

Figure 2.16: The convolution process of characteristic functions is illustrated above. One of the functions' moves is multiplied by the second, and the average of the product is the result.

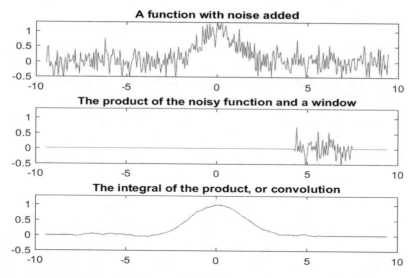

Figure 2.17: The idea of using convolution to create a moving average, and denoise a noisy function is illustrated above. The noisy function is multiplied by a simple averaging function, which is 1 on a region, and then an average is done.

2. Prove that $f(t) = 2\pi - |t|$ on $[-2\pi, 2\pi]$ is the convolution of $\chi_\pi(t)$ against itself, or $\chi_\pi * \chi_\pi(t)$.

3. Calculate the Fourier Series for the hat function $f(t) = 2\pi - |t|$ on $[-2\pi, 2\pi]$ in two ways: (a) by calculating its Fourier Series directly. (b) by calculating the Fourier Series for $\chi_\pi(t)$ and using the convolution theorem.

4. **Challenging** Consider the two functions,

$$f(t) = \begin{cases} \cos(x) & \text{if} \quad |t| < \pi/2 \\ 0 & \text{if} \quad |t| \geq \pi/2 \end{cases}. \tag{2.58}$$

and

$$g(t) = \begin{cases} 0 & \text{if} \quad |t| > .1 \\ -10 & \text{if} \quad t \in [-.1, 0] \\ 10 & \text{if} \quad t \in [0, .1] \end{cases}. \tag{2.59}$$

(a) Compute the Fourier Transforms of both functions on $[-\pi, \pi]$. (b) Compute the convolution $f * g$ via the convolution theorem. (c) Plot the first 25 terms of the result. (d) Is the result similar to the derivative of f? (e) Why do you think this is?

2.9 Chapter Project:

1. Compute the Fourier Series for the function $f(t) = t$ which converges a) on $[-\pi, \pi]$ and b) on $[-2\pi, 2\pi]$. Plot the approximations using 5,10, and 15 terms on $[-4\pi, 4\pi]$. These are two different series and should look different.

2. Figure out the necessary cosine, and sine terms and the appropriate coefficient formulas to approximate a function on the interval $[1, 3]$.

3. Using the result from Problem 2 above, calculate the first few terms of the expansion of $f(t) = t$ on $[1, 3]$. Plot this result on $[-1, 5]$.

4. Verify that the Fourier isometry holds on $[-\pi, \pi]$ for $f(t) = t$. To do this, a) calculate the coefficients of the orthogonal Fourier Series from the orthogonal series representation, b) calculate the sum of the squared coefficients, and c) Calculate the norm of the function as $\int_{-\pi}^{\pi} |f(t)|^2 dt$. They must be equal. How many terms in the Fourier Series are necessary to have the isometry be under 5%? How many until you are under 3%, or 1%?

5. **Gibbs ringing:** Calculate the Fourier Series for $\chi_{1/2}(t)$ on $[-1, 1]$, and plot the first 50 terms. Find the maximum value of the series with 30, 40, 50, and 100 terms. Guess at the limit of this maximum value. Find the location of the maximum value on the right of 0, with 30, 40, 50, and 100 terms. Where is the maximum value headed? Finally, estimate the squared error $\int_{-1}^{1} |\chi_{1/2}(t) - S_n(\chi_{1/2})(t)|^2 dt$ and the series on $[-1, 1]$ with the same number of terms (do this numerically).

6. **Sine and cosine series:** Calculate the sine and cosine series for $f(t) = t$ on $[0, 1]$. Plot the first 30 terms of these series on $[-3, 3]$. Estimate the error between both series and the function on $[0, 1]$ after 30 terms. How many more terms of the sine series are necessary to achieve the same error as was achieved with the cosine series and 30 terms?

7. **Sine and cosine series:** Plot the first 30 terms of the derivatives of both the sine and cosine series in the above problem. What do you observe? Do both of them converge?

8. **Convolution:** Calculate the Fourier Series for the hat function $f(t) = 2\pi - |t|$ on $[-2\pi, 2\pi]$ in two ways. a) by calculating its Fourier Series directly. b) by calculating the Fourier Series for $\chi_{\pi}(t)$ and using the convolution theorem. They must be equal! Use trig identities. . .

2.10   Summary of Expansions:

- **Fourier expansion for** $f(t) \in L^2[-\pi, \pi]$:

  Basic Formula:

  $$f(t) = \frac{a_0}{2} + \sum_{k=0}^{\infty} a_k \cos(kt) + b_k \sin(kt),$$

  where

  $$a_k = \frac{1}{\pi} \int_{-\pi}^{\pi} f(t) \cos(kt) dt \text{ and } b_k = \frac{1}{\pi} \int_{-\pi}^{\pi} f(t) \sin(kt) dt.$$

  Orthonormal Expansion:

  $$f(t) = a_0 \frac{1}{\sqrt{2\pi}} + \sum_{k=0}^{\infty} a_k \frac{\cos(kt)}{\sqrt{\pi}} + b_k \frac{\sin(kt)}{\sqrt{\pi}},$$

  where $a_0 = 1/\sqrt{2\pi} \int_{-\pi}^{\pi} f(t) dt$ and for $k \geq 1$

  $$a_k = \int_{-\pi}^{\pi} f(t) \frac{\cos(kt)}{\sqrt{\pi}} dt \text{ and } b_k = \int_{-\pi}^{\pi} g(t) \frac{\sin(kt)}{\sqrt{\pi}} dt.$$

- **Fourier expansion for** $f(t) \in L^2[-T, T]$:

  Basic Formula

  $$f(t) = \frac{a_0}{2} + \frac{1}{T} \sum_{k=0}^{\infty} a_k \cos\left(\frac{k\pi t}{T}\right) + b_k \sin\left(\frac{k\pi t}{T}\right),$$

  where

  $$a_k = \int_{-T}^{T} f(t) \cos\left(\frac{k\pi t}{T}\right) dt \text{ and } b_k = \int_{-T}^{T} f(t) \sin\left(\frac{k\pi t}{T}\right) dt.$$

  Orthonormal Expansion:

  $$f(t) = a_0 \frac{1}{\sqrt{2T}} + \sum_{k=0}^{\infty} a_k \frac{\cos(\frac{k\pi t}{T})}{\sqrt{T}} + b_k \frac{\sin(\frac{k\pi t}{T})}{\sqrt{T}},$$

  where $a_0 = 1/\sqrt{2T} \int_{-\pi}^{\pi} f(t) dt$ and for $k \geq 1$

$$a_k = \int_{-T}^{T} f(t) \frac{\cos(\frac{k\pi t}{T})}{\sqrt{T}} dt \text{ and } b_k = \int_{-T}^{T} f(t) \frac{\sin(\frac{k\pi t}{T})}{\sqrt{T}} dt.$$

- **Fourier expansion for** $f(t) \in L^2[a, b]$: Let $m = (a+b)/2$, and $L = (b-a)/2$.

$$f(t) = \frac{a_0}{2} + \frac{1}{L} \sum_{k=0}^{\infty} a_k \cos\left(\frac{k\pi(t-m)}{L}\right) + b_k \sin\left(\frac{k\pi(t-m)}{L}\right),$$

where

$$a_k = \int_a^b f(t) \cos\left(\frac{k\pi(t-m)}{L}\right) dt \text{ and } b_k = \int_a^b g(t) \sin\left(\frac{k\pi(t-m)}{L}\right) dt.$$

Orthonormal Expansion:

$$f(t) = a_0 \frac{1}{\sqrt{2L}} + \sum_{k=0}^{\infty} a_k \frac{\cos\left(\frac{k\pi(t-m)}{L}\right)}{\sqrt{L}} + b_k \frac{\sin\left(\frac{k\pi(t-m)}{L}\right)}{\sqrt{L}},$$

where $a_0 = 1/\sqrt{2L} \int_a^b f(t) dt$ and for $k \geq 1$

$$a_k = \int_a^b f(t) \frac{\cos\left(\frac{k\pi t}{L}\right)}{\sqrt{L}} dt \text{ and } b_k = \int_a^b f(t) \frac{\sin\left(\frac{k\pi t}{L}\right)}{\sqrt{L}} dt.$$

- **Cosine and Sine expansions for** $f(t) \in L^2[0, \pi]$:

Basic Formulas:

$$f(t) = \frac{a_0}{2} + \sum_{k=0}^{\infty} a_k \cos(kt),$$

and

$$f(t) = \sum_{k=0}^{\infty} b_k \sin(kt),$$

where

$$a_k = \frac{2}{\pi} \int_0^{\pi} f(t) \cos(kt) dt \text{ and } b_k = \frac{2}{\pi} \int_0^{\pi} f(t) \sin(kt) dt.$$

Orthonormal Expansion:

$$f(t) = a_0 \frac{1}{\sqrt{\pi}} + \sum_{k=0}^{\infty} a_k \sqrt{\frac{2}{\pi}} \cos(kt)$$

and

$$f(t) = \sum_{k=0}^{\infty} b_k \sqrt{\frac{2}{\pi}} \sin(kt),$$

where $a_0 = 1/\sqrt{\pi} \int_0^{\pi} f(t)dt$ and for $k \geq 1$

$$a_k = \int_0^{\pi} f(t) \sqrt{\frac{2}{\pi}} \cos(kt)dt \text{ and } b_k = \int_0^{\pi} f(t) \sqrt{\frac{2}{\pi}} \sin(kt)dt.$$

Chapter 3
The Discrete Fourier Transform

Fourier Series is a way to represent a function $f(t) \in L^2[a, b]$ of a continuous variable t with a countable number of coefficients c_k. Oftentimes, however, we are interested in representing a finite number of data points $\{f(t_k)\}_{k=0}^{N-1}$ which probably come as samples of a function of a continuous variable t. We may not have enough information to represent the original function, but we would still like to know what its Fourier Transform looks like. We have three primary goals for our discrete Fourier Analysis: (1) having an accurate representation, (2) being able to calculate the representation quickly and easily, and (3) knowing what the coefficients of that representation represent. We will now try to accomplish these goals.

In this section, we will concentrate entirely on data or a string of numbers which are believed to be part of some process. From the viewpoint of Linear Algebra, these are merely vectors. We generally think of this as an approximation to the real underlying continuous functions. Let us assume that the data points are t_k where $k = 0, 1, 2, 3, \ldots N - 1$. It really does not matter how we label them, or whether t_k is in seconds, years, or nanoseconds. We can call the first time we measure time 0. The one thing we will insist on is that they are spaced uniformly. The question becomes, "How do we construct a Fourier Transform for a finite data set?" The answer is simple. We do it the same way we did it with a function of the continuous variable t.

We begin by making sure that the first sine and cosine have exactly one cycle from beginning to end. Then, we make sure things are adjusted so that we have "enough" sines and cosines. Let us do this. We want ht_k to go from 0 to 2π, while k goes from 0 to $N - 1$ so that the first sine and cosine will have exactly one cycle. We want $\sin(hN) = 0$, or we want the fictional $N'th$ step to be 2π. This is something that happens when you discretize. Thus if we let $hN = 2\pi$, we find that $h = 2\pi/N$, or we want our basis functions to include

$$\sin\left(\frac{2\pi k}{N}\right),$$

© Springer Science+Business Media, LLC 2017
T. Olson, *Applied Fourier Analysis*,
https://doi.org/10.1007/978-1-4939-7393-4_3

and a similar set of cosines. We would like N such functions so we consider the functions

$$\sin\left(\frac{2\pi nk}{N}\right),$$

where $n = 0 \ldots N - 1$. We do not want to have separate cosine and sine coefficients, so for convenience, we will go immediately to complex notation. These will not be the Fourier functions, but rather the Fourier vectors

$$\{f_n\} = \left\{\exp\left(\frac{2\pi ink}{N}\right)\right\}_{k=0}^{N-1}, \tag{3.1}$$

where we have N vectors $n = 0, 1, \ldots N - 1$.

The vectors above have a beautiful mathematical basis. Namely, each vector above has a natural expression in terms of the N'th complex roots of unity. What do we mean by this? Simply stated, all of the above vectors come from solutions to $x^N = 1$, or if we allow x to be complex, we call this $z^N = 1$.

The 2'nd roots of unity are simple. They are 1 and -1. The solutions of $z^3 = 1$ are literally more complex. They are still simply expressed as $\exp((2\pi ik)/3)$, however. When $k = 0$, you have 1. When $k = 1$, you have $\exp(2\pi i/3)$. Note that $\exp(2\pi i/3)^3 = \exp(2\pi i) = 1$ so it is also a solution. When $k = 2$, you have $\exp(4\pi/3i)$ and $\exp(4\pi/3i)^3 = \exp(4\pi i) = 1$, and thus, you have all three solutions to the 3'rd degree polynomial $z^3 = 1$. Notice that the root when $k = 2$ is the root when $k = 1$ squared, or $\exp(2\pi/3) = \exp(\pi/3)^2$. The third root is also the complex conjugate of the second root. This will continue to be a recurring theme. There is a great deal of structure in these solutions or roots. We will now define our notation for these roots.

Definition 3.0.1 (N'th roots of unity) *The solution to the problem $z^N = 1$, given by $z = \exp(2\pi i/N)$ is called the N'th primitive root of Unity. We will denote it by w_N. Often, however, we will not utilize the subscript N, since the dimension N will be obvious in context.*

3.1   The Fourier Matrix

Now if we return to the definition of the Fourier vectors 3.1, we have

$$\{f_n\} = \left\{\exp\left(\frac{2\pi ink}{N}\right)\right\}_{k=0}^{N-1}$$

$$= \left\{w^{kn}\right\}_{k=0}^{N-1}, \tag{3.2}$$

where once again we have N vectors, $n = 0, 1, 2, \ldots N - 1$. Now writing one of these vectors out, we get

$$f_n = [1, w^n, w^{2n}, w^{3n}, w^{4n}, \ldots w^{(N-1)n}]^t.$$

The Fourier matrix is thus

$$F = \frac{1}{\sqrt{N}} \begin{pmatrix} - - f_0 - - \\ - - f_1 - - \\ - - f_2 - - \\ \vdots \\ - - f_{N-1} - - \end{pmatrix}$$

where the row vectors of the matrix are exactly the Fourier vectors that we had designed. This implies that the matrix multiplication $\vec{c} = F\vec{x}$ is a vector of dot products against our Fourier vectors. We now explicitly write the complete matrix when $N = 4$

$$F_4 = \frac{1}{2} \begin{pmatrix} 1 & 1 & 1 & 1 \\ 1 & w & w^2 & w^3 \\ 1 & w^2 & w^4 & w^6 \\ 1 & w^3 & w^6 & w^9 \end{pmatrix}.$$

The general NxN Fourier matrix is then given by

$$F_N = \frac{1}{\sqrt{N}} \begin{pmatrix} 1 & 1 & 1 & 1 & . & 1 \\ 1 & w & w^2 & w^3 & . & w^{N-1} \\ 1 & w^2 & w^4 & w^6 & . & w^{2(N-1)} \\ 1 & w^3 & w^6 & w^9 & . & w^{3(N-1)} \\ 1 & . & . & . & & . \\ 1 & w^{N-1} & w^{2(N-1)} & w^{3(N-1)} & . & w^{(N-1)(N-1)} \end{pmatrix} \tag{3.3}$$

where the fraction is a normalization factor which will be explained later. Notice that the matrix is symmetric. This is very important.

3.1.1 Orthogonality of the Fourier Matrix

We emphasized earlier that one of the advantages of the sines and cosines is that they are naturally orthogonal on the appropriate intervals. This is contrasted against polynomials, which must be made orthogonal. There are certainly orthogonal polynomials. One of the things that set Fourier Series apart from orthogonal polynomials is that the uniformly sampled discrete vectors associated with Fourier Series functions are also orthogonal. Certain polynomials on certain points are also orthogonal (Chebyshev polynomials on the Chebyshev nodes), but these are merely the projections of Fourier Series onto the line.

Let us remember that we are using the complex inner product, which specifies that $\langle \{a_k\}, \{b_l\} \rangle = \sum a_k \bar{b}_k$, with $\bar{b}$ being the complex conjugate $\overline{\alpha + i\beta} = \alpha - i\beta$. First, consider

$$\langle f_n, f_n \rangle = \sum_{k=0}^{N-1} w^{kn}\overline{w}^{kn} = \sum_{0}^{N-1} (w\overline{w})^{kn} = \sum_{0}^{N-1} 1 = N.$$

To understand why all of the terms are 1, remember that $w^k = exp(i2\pi kn/N)$ and $\overline{w}^k = exp(-i2\pi kn/N)$ so $w^k\overline{w}^k = 1$. This also explains why we have the $1/\sqrt{N}$ factor in front of the matrix. Putting this in front of the f_n's makes them normalized, and we will see that it makes the matrix (almost) orthonormal. Another way to remember this is that any complex number whose absolute value is 1 has the relation $w\overline{w} = 1$.

Now consider

$$\langle f_{n1}, f_{n2} \rangle = \sum_{k=0}^{N-1} w^{k(n1)}\overline{w}^{k(n2)} = \sum_{k=0}^{N-1} w^{k(n1-n2)} = \sum_{k=0}^{N-1} w^{kq},$$

where $q = n1 - n2$. Note that any multiple of the primitive N'th root of the unity w is also an N'th root of unity. The above geometric sum is then equal to

$$\frac{1 - w^{qN}}{1 - w^q}.$$

Since w^q is an N'th root of unity, the denominator $1 - w^{qN} = 0$. Furthermore if $q = n_1 - n_2 \neq 0$, then q is not a multiple of N so the numerator $1 - w^q \neq 0$. Thus, we have

$$\langle f_{n1}, f_{n2} \rangle = \frac{1 - w^{qN}}{1 - w^q} = 0$$

whenever $n_1 \neq n_2$ which proves orthogonality for each of the vectors.

We have shown that the individual row vectors are orthogonal. With a real matrix, the orthogonality of the vectors would mean the transpose of the matrix is the inverse or $F^t = F^{-1}$. We must remember that these are complex vectors, so the inner product is $\langle f_k, f_j \rangle = \sum_{m=0}^{N-1} f_k(m)\hat{f}_j(m)$. Therefore, we must introduce the conjugate transpose of the matrix, which is $F^* = \overline{F}^t$. In other words, we take the conjugate of each of the matrix components and then transpose the matrix.

This tells us that $F^*F = I$ or that the conjugate transpose of the Fourier matrix is its inverse. This leads to revisiting the isometry in another setting.

Theorem 3.1.1 (Fourier Isometry from $\mathbb{C}^n$ to $\mathbb{C}^n$) *The Discrete Fourier Transform is an isometry from $\mathbb{C}^n$ to $\mathbb{C}^n$. Specifically, $\|F\vec{x}\| = \|\vec{x}\|$.*

Proof: The proof is straightforward and will be done in one line. One must just remember that $\|\vec{a}\|^2 = \vec{a} \cdot \vec{a}$ and that $\vec{a} \cdot \vec{a} = \vec{a}^t \vec{a}$. The dot product is a matrix multiplication between rows on the left and columns on the right. Now, we have

$$\|F\vec{x}\|^2 = \overline{\vec{x}^t F^t} F\vec{x} = \overline{\vec{x}^t} \overline{F^t} F\vec{x} = \overline{\vec{x}^t} I \vec{x} = \overline{\vec{x}^t} \vec{x} = \|\vec{x}\|^2.$$

3.2 The Complex N'th roots of unity and their structure

The Fourier matrix F_N introduced above has a great deal of structure involved in it. It is created by powers of the N'th roots of unity. Mathematically, these roots of unity have very special properties, which we need to understand. The 2'nd roots are merely 1 and -1. The 4'th roots of unity are more interesting, namely $1, i, -1, -i$. The structure of the N'th roots of unity is very simple and at the same time very useful. At this time, let us restate the definition of the primitive root and state a number of the properties which are associated with the N'th roots of unity.

Definition 3.2.1 (N'th roots of unity) *The solution to the problem $z^N = 1$, given by $z = \exp(2\pi i/N)$, is called the N'th primitive root of unity. We will denote it by w_N. We will oftentimes suppress the subscript N, since the dimension N will be obvious in context.*

Let us now state a number of the properties and which the N'th roots of unity possess. All of the rules below can be verified very simply. We leave these as an exercise. While they are simple, they aid greatly in understanding the Fourier matrix and the Discrete Fourier Transform and their relationship to Fourier Series. We will try to explain these relationships more clearly in the next section.

3.2.1 Problems and Exercises:

1. All N of the solutions to $z^N = 1$ are generated powers of w_N. In other words, all of the N'th roots of unity are given by $1, w_N, w_N^2, w_N^3, w_N^4, \ldots,$ w_N^{N-1}, or more cleanly, $\{w_N^k\}_{k=0}^{N-1}$. The sixteens roots of unity are shown on the unit circle in the complex plane in Figure 3.2.

2. Raising a primitive root of unity to higher powers $k > 1$, or w_N^k, results in geometrically moving these respective roots counterclockwise around the unit circle.

3. All N of the solutions to $z^N = 1$ are generated powers of w_N^{-1}. Thus, all of the N'th roots of unity are given by $1, w_N^{-1}, w_N^{-2}, w_N^{-3}, w_N^{-4}, \ldots, w_N^{-(N-1)}$, or more cleanly, $\{w_N^{-k}\}_{k=0}^{N-1}$. The sixteenth roots of unity are shown on the unit circle in the complex plane in Figure 3.2.

4. Raising the root of unity w_N^{-1} to higher powers k, or w_N^{-k}, results in geometrically moving these respective roots clockwise around the unity circle.

5. If N is divisible by 2, then the (N/2)th primive root of unity will be given by $w_N^2 = w_{N/2}$. This is also illustrated in Figure 3.2 where $w_8 = w_{16}^2$.

6. If $N = N_1 N_2$ where N_1 and N_2 are integers, then $w_N^{N_1} = w_{N_2}$ and $w_N^{N_2} = w_{N_1}$.

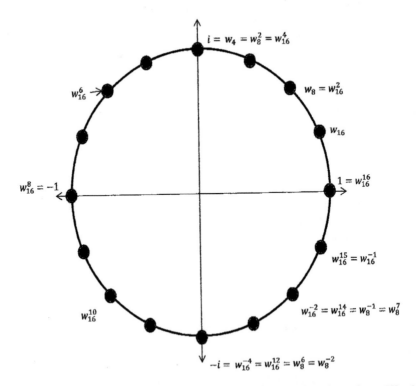

Figure 3.1: We illustrate the 16'th roots of unity in the complex plane above. We also point out some of the relationships with the 8'th roots of unity.

7. The root w_N^{-k} is equal to the root w_N^{N-k} or $w_N^{-k} = w_N^{N-k}$.

8. The complex conjugate of w_N^k is w_N^{N-k}. The arithmetic on this is simple, but not completely intuitive to begin with. This is also illustrated in Figure 3.2.

3.2.2   The roots of unity and frequency

Algebraically, it is fairly easy to see that $w_N^{(N-1)} = w_N^N w_N^{-1} = w_N^{-1}$. Therefore, it makes sense that w_N and w_N^{N-1} are complex conjugates of each other. This is further clarified when we remember that

$$z^{-1} = \frac{1}{z} = \frac{\overline{z}}{z\overline{z}} = \frac{\overline{z}}{|z|^2}.$$

It follows that if $|z| = 1$, as is the case with the roots of unity, then $z^{-1} = \overline{z}$. We have now shown that $\overline{w}_N = w_N^{-1} = w_N^{N-1}$.

While this algebraic argument with complex exponentials is easy, it seems harder to understand why the points $w_N^k = \cos(2k\pi/N) + i\sin(2k\pi/N)$ have the complex conjugate points $w_N^{k(N-1)} = \cos(k(N-1)2\pi/N) + i\sin(k(N-1)2\pi/N)$. If we think of this as a vector indexed by k, why are these vectors complex conjugates of each other? In the case where $N = 16$, it would seem that $\cos(k2\pi/16)$ would go through one cycle as $k = 0, 1, 2, \ldots 16$, while $\cos(k2\pi 15/16)$ would go from 0 to $256/16\pi$ as $k = 0, 1, 2 \ldots 16$, thus implying 16 cycles. This false illusion is caused by the deceiving nature of the periodicity of the sines and cosines. The value of $\cos(2\pi 15/16)$ is one stop short on our discretized grid of the circle from 2π. Now, $\cos(2\pi * (2)15/16) = \cos(2\pi 30/16)$ is actually two spots short of 4π in our discretized grid. Similarly, the next step will be 3 steps short of 6π. Thus, the periodicity has $\cos(k2\pi 15/16)$ skipping between cycles of the infinite cosine, but always losing one spot.

This is exactly why $\cos(k2\pi/16) = \cos(-k2\pi 15/16)$. Similarly, $\sin(k2\pi/16) = -\sin(k2\pi 15/16)$, or $\sin(k2\pi/N) = -\sin(k2\pi(N-1)/N)$. Thus, the cosines agree and the sines are opposite, showing that they are complex conjugates after all.

The roots w_N^m where $m < N/2$

It is enlightening to return to Figure 3.2. If you start out with w_{16}^2, then $(w_{16}^2)^2 = w_{16}^4$. Each time k advances, we move forward two steps on the unit circle. If we start with w_N^m, then we move forward m spots on the circle with each additional k. This is accurate and correct for all of the roots of unity on the top half of the unit circle. Thus, the m'th row of the Fourier matrix F_N, given by $\{w_N^{mk}\}_{k=0}^{N-1}$ will go through exactly m cycles as k goes from 0 to $N-1$, if $m < N/2$

The roots w_N^m where $m > N/2$

This type of approach is deceiving for roots of unity on the bottom half of the unit circle in Figure 3.2 or when $M > N/2$. If we consider w_{16}^{15}, then $w_{16}^{2(15)} = w_{16}^{30}$ would seem to be far way from w_{16}^{15}. Because of the periodicity, however, $w_{16}^{30} = w_{16}^{14}$ which is one spot away from w_{16}^{15}. While this is correct, it is much easier to see that $w_{16}^{15} = w_{16}^{-1}$; therefore, $(w_{16}^{15})^2 = (w_{16}^{-1})^2 = w_{16}^{-2}$. Thus, w_{16}^{15} advances only one spot with each k and therefore will have only one cycle, moving clockwise rather than counterclockwise around the unit circle. This is similarly true for any $m > N/2$, where $w_N^{km} = w_N^{k(m-N))}$. Remember that $m - N$ is negative, and $m - N > -N/2$ so these roots will move in a clockwise manner moving $N - m$ steps in the clockwise direction for each increase in k. They will have exactly $N - m$ cycles just as their conjugates on the upper half of the circle.

Let us now return to the Fourier matrix with these discussions in mind.

3.2.3 Making the Fourier Matrix Understandable

We discussed above that the periodicities and structure of the Fourier matrix sometimes make its output mysterious and can lead to incorrect understanding. The discussion above helps. We will now rewrite at least one Fourier matrix, in a much more understandable form.

We will pick F_8 as our example, because it is big enough to illustrate our point and small enough to fit on one page. The standard matrix where we let $w_8 = w$ (deleting the subscript for clarity) is

$$F_8 = \frac{1}{\sqrt{8}} \begin{pmatrix} 1 & 1 & 1 & 1 & 1 & 1 & 1 & 1 \\ 1 & w & w^2 & w^3 & w^4 & w^5 & w^6 & w^7 \\ 1 & w^2 & w^4 & w^6 & w^8 & w^{10} & w^{12} & w^{14} \\ 1 & w^3 & w^6 & w^9 & w^{12} & w^{15} & w^{18} & w^{21} \\ 1 & w^4 & w^8 & w^{12} & w^{16} & w^{20} & w^{24} & w^{28} \\ 1 & w^5 & w^{10} & w^{15} & w^{20} & w^{25} & w^{30} & w^{35} \\ 1 & w^6 & w^{12} & w^{18} & w^{24} & w^{30} & w^{36} & w^{42} \\ 1 & w^7 & w^{14} & w^{21} & w^{28} & w^{35} & w^{42} & w^{49} \end{pmatrix}. \qquad (3.4)$$

From our discussions above, we know that the complex conjugate pairs for $w_8 = w$ in this case are: $\{w, w^7\}, \{w^2, w^6\}, \{w^3, w^5\}$. There are no corresponding conjugate pairs for $w^0 = 1$, or $w^4 = -1$, since they are both real. While this allows us to understand the 1'st and 7'th rows, the 2'nd and 6'th, and the 3'rd and 5'th rows are associated, it still makes one think. If, however, we write $w^7 = w^{-1}$, $w^6 = w^{-2}$, $w^5 = w^{-3}$, and substitute $w^4 = -1$ into the matrix, we get the following matrix

$$
F_8 = \frac{1}{\sqrt{8}}
\begin{pmatrix}
1 & 1 & 1 & 1 & 1 & 1 & 1 & 1 \\
1 & w & w^2 & w^3 & w^4 & w^5 & w^6 & w^7 \\
1 & w^2 & w^4 & w^6 & w^8 & w^{10} & w^{12} & w^{14} \\
1 & w^3 & w^6 & w^9 & w^{12} & w^{15} & w^{18} & w^{21} \\
1 & -1 & 1 & -1 & 1 & -1 & 1 & -1 \\
1 & w^{-3} & w^{-6} & w^{-9} & w^{-12} & w^{-15} & w^{-18} & w^{-21} \\
1 & w^{-2} & w^{-4} & w^{-6} & w^{-8} & w^{-10} & w^{-12} & w^{-14} \\
1 & w^{-1} & w^{-2} & w^{-3} & w^{-4} & w^{-5} & w^{-6} & w^{-7}
\end{pmatrix}
\tag{3.5}
$$

This is exactly the matrix written above, but is much more understandable. Now, it is clear that the 2'nd and 8'th rows are paired. Multiplication of this matrix against a vector, especially a real-valued vector, should leave very structured results. We leave these as exercises.

While this discussion has cleared up some of the misunderstandings of the Fourier matrix, let us now introduce a shift operator which makes it easier to manipulate the output of the matrix.

Reorganizing the coefficients

Often, this is not an easy format to visualize or manipulate, so one wants to reorganize the coefficients $\hat{f} = \{c_k\}_{k=0}^{N-1}$ in the form $\hat{f}_{shift}\{\{c_k\}_{N/2+1}^{N-1},$ $\{c_k\}_{k=0}^{N/2}\}$. This will place the lowest (constant) Fourier coefficient at the middle of the vector, and the higher coefficients will be symmetrically aligned around this. This is accomplished in MATLAB by the simple command fft-shift. Other packages generally have similar commands. For clarity in this book, we will define the shift operator.

Definition 3.2.2 (Shift Operator) *Given a vector* $\{x_n\}_{n=0}^{N-1}$, *we define the shift operator* S_k *as*

$$
S_k(\{x_n\}_{n=0}^{N-1}) = \{\{x_n\}_{n=k+1}^{N-1}, \{x_n\}_{n=0}^{k}\}. \tag{3.6}
$$

Note that this is in essence a rotational shift, where the last becomes first, etc. Secondly, the indexing of the vector does not matter, as long as the $k+1$'st entry becomes the first.

It is important to note one relationship that $S_k(S_n) = I$ if $k + n = N$. Oftentimes in this book, we are only interested in $S_{N/2}$ when a vector has N entries. We illustrate the use of the shift operator in Figure 3.2.

3.2.4   Problems and Exercises:

1. Let $\vec{f} = \{f_k\}_{k=0}^{N-1}$ be a real-valued vector. Show that $\hat{f} = F\vec{f}$ is conjugate symmetric, in the sense that $Re(\hat{f}_k) = Re(\hat{f}_{N-k})$, and $Im(\hat{f}_k) = -Im(\hat{f}_{N-k})$.

2. Suppose that $\vec{f}$ is a positive vector. Show that $\hat{f}_0$ must be larger than all other elements $\hat{f}_k$.

3. Suppose that $\vec{f} = \{f_k\}_{k=0}^{N-1}$ is symmetric in the sense that $f_k = f_{N-k}$ for $0 \le k < N/2$. Prove that $\hat{f}$ is symmetric and real.

4. Suppose that $f = \{f_k\}_{k=0}^{N-1}$ is antisymmetric in the sense that $f_k = -f_{N-k}$ for $0 \le k < N/2$. Similar to Problem 3, what can you prove about $\hat{f}$?

5. If $\hat{f}$ is conjugate symmetric, in the sense that $Re(\hat{f}_k) = Re(\hat{f}_{N-k})$, and $Im(\hat{f}_k) = -Im(\hat{f}_{N-k})$, for $0 \le k < N/2$. Is f necessarily real? Can you prove this?

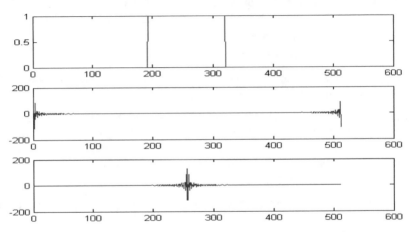

Figure 3.2: We illustrate a discrete characteristic function χ and its Discrete Fourier Transform. The discrete characteristic is shown first. The second illustration is its Discrete Fourier Transform as it would be output from the standard Fourier matrix. The third is the shifted version, with $S_{N/2}$.

3.2.5 Group structure of the roots of unity

We have observed that there is a great deal of structure involved with the Fourier matrix. Mathematically, this structure can be described by saying that the roots of unity form a group under standard multiplication. To begin with, if we start with the primitive N^{th} root w, we can generate all of the other roots of unity by repeated multiplication. Specifically, starting with w, we generate $w^2, w^3, w^4, \ldots w^{N-1}, 1$ by simply multiplying by w. Under standard rules of algebra, this is a group. Specifically multiplication of any element by any other element, $w^n w^m = w^{nm}$, yields another root of unity.

There is an identity element 1, and every element has its inverse, which is $w^k = w^{N-k} = w^{-k} = \overline{w}^k$. For inclusiveness, we state the definition of a group.

Definition 3.2.3 (Group) *We say that a collection of elements $\{g\}$ is a group G if there is an associated operation $\cdot$ such that $g_m \cdot g_n \in G$ for all elements of G. In addition: (i) the operation $\cdot$ must be associative, i.e., $a \cdot (b \cdot c) = (a \cdot b) \cdot c$; (ii) there must be an identity element e such that $e \cdot g = g \cdot e$ for all $g \in G$; and (iii) there must be an inverse element of any element of G, i.e., for every $g \in G$ there must be an h such that $g \cdot h = h \cdot g = e$.*

We need to introduce another concept which fits with the group structure of the roots of unity. For a particular case, suppose that we are dealing with the 7th roots of unity. Then, we would have that $w^9 = w^2 w^7 = w^2(1) = w^2$. Thus, we would like to associate 1 with 8, 2 with 9, 3 with 10, and 4 with 11. This is called modular arithmetic.

Definition 3.2.4 (Modular Arithmetic) *Two integers m and n are considered to be equal, modulo N, if $m - n = Nk$, where k is any integer. In this case, we may write $m \equiv n (mod N)$.*

This brings a more basic or perhaps fundamental concept into play. This is very similar to the isometry of Chapter 2, which means same metric, in both spaces. There we had that

$$\|f(t)\|_2 = \|g(t)\|_2$$

implies that

$$\|\mathcal{F}(f)\|_2 = \|\mathcal{F}(g)\|_2,$$

or that the Fourier Series operator $\{\mathcal{F}\}$ maintains distances, or we have equal (iso) metrics, under the mapping.

We need to introduce a similar idea, which is a homomorphism. This is almost identical in that iso, or one, and homo, meaning same, come from the same Latin meanings. Isometries maintain distance. Homomorphisms maintain the appropriate arithmetic. Thus, let us define the following.

Definition 3.2.5 (Homomorphism) *A map H between two groups, G_1 and G_2 is said to be a homomorphism if whenever $a, b \in G_1$, it follows that 1) $H(a), H(b) \in G_2$, and $H(a \cdot b) = H(a) \cdot H(b)$, where $\cdot$ is the operation in the respective group.*

Thus, a homomorphism is a mapping that maintains the group arithmetic through the mapping. Homomorphisms are very useful for understanding the Discrete Fourier Transform (DFT). Primarily from the above discussion, let us define two groups.

Definition 3.2.6 (The group) $\mathbb{Z}_N$) *Let $G = \mathbb{Z}_N$ be set of integers $0, 1, 2, 3$, ...$N - 1$. For any two elements of $x, y \in \mathbb{Z}_N$ define the binary operation $a \cdot b = (a + b) Mod(N)$, where modulo arithmetic is defined as above.*

The fact that $G = \mathbb{Z}_N$ is a group is left as an exercise.

Definition 3.2.7 (N^{th} Roots of Unity) *Let $G_2 = \{w_N^k\}_{k=0}^{N-1}$ where $w_N = \exp(2\pi i/N)$ be the solutions to the equation $z^N = 1$ in the complex plane. Let the binary operation $\cdot$ be multiplication in the complex plane. Then, G_2 is a group with this operation.*

The product of any two N^{th} roots of unity is an Nth root of unity. The fact that G_2 is a group is also left as an exercise. What we want to establish is the following.

Theorem 3.2.1 *The groups $\mathbb{Z}_N$ and G_2 are homomorphic to each other under the simple map $k \rightarrow w_N^k$.*

The proof of this is left as an exercise. Quite simply, multiplying exponentials means adding their exponents. Thus, multiplication in one group is addition in the other.

3.2.6   Subgroups and Coset Representations

We will now introduce the idea of a subgroup.

Definition 3.2.8 (Subgroup *A proper subgroup of a finite group is a subset which remains closed with the multiplication of the group, but which does not contain all members of the group.*

For the purposes of this book, we are primarily concerned with the subgroups of $\mathbb{Z}_N$ and therefore the subgroups of the N'th roots of unity. It is fairly easy to see that the group $\mathbb{Z}_6 = \{0, 1, 2, 3, 4, 5\}$ has two proper subgroups. The first is $\{0, 3\}$, which is closed under addition, since $3 + 3 = 0$. The second is $\{0, 2, 4\}$, which is also a proper subgroup. It is not a coincidence that $N = 6 = 2 \cdot 3$, and these two subgroups are those generated by 2 and 3. If N has a primary factorization $N = p_1 p_2 p_3 ... p_n$, then $\mathbb{Z}_N$ will have subgroups generated by all of these prime factors.

We will use the notation $< k >$ to denote the subgroup generated by the action of the group addition utilizing the original element k. If we are in $\mathbb{Z}_N$, then this will compromise the addition of k repeatedly, i.e., $k, k + k, k + k + k, ...$ or more simply $k, 2k, 3k, 4k, ...$ Note that if k is a divisor of N, this will be a proper subgroup. If k and N are relatively prime, however, then the subgroup generated by k or $< k >$ will be a whole group and will not be a proper subgroup. These ideas are left as exercises.

We will state a theorem which is fundamental to understanding the Discrete Fourier Transform. This involves the N'th roots of unity and is stated quite simply.

Theorem 3.2.2 *Let $G(N)$ be the group formed by the Nth roots of unity. Assume that we can factor N into two integers, $N = n_1 n_2$. Then, the group $G(n_1)$ formed by the n_1th roots of unity and the group $G(n_2)$ formed by the n_2th roots of unity are both proper subgroups of $G(N)$. Stating this another way the n_1th roots of unity are also N'th roots of unity.*

This proof is left as an exercise.

A concept which will be fundamental to the Fast Fourier Transform (FFT), is that of coset representation. In the case of $\mathbb{Z}_N$, this is fairly easy and we will define this now with the idea carrying over to the roots of Unity.

Definition 3.2.9 (Coset Representation for $\mathbb{Z}_N$) *Let us assume that N can be factored into two integers such that $N = mn$. Then the subgroups $< m >$ and $< n >$ will be proper subgroups. Moreover for any k in $\mathbb{Z}_N$ we can write $k = j_1 n + j_2$ where $j_1 \in \mathbb{Z}_m$ and $j_2 \in \mathbb{Z}_n$. Another way to write this is to represent the elements of $\mathbb{Z}_N$ as the collection of cosets beginning with the subgroup generated by n, i.e., $< n >$, and then adding the elements $j = 0, 1, ..., m - 1$. Thus, we have a unique set of cosets, $< n > + j$, where $j = 0, 1, .., m - 1$, and another unique set of cosets $< m > + k$, where $k = 0, 1, ..n - 1$.*

Let us use $\mathbb{Z}_6$ to illustrate this. Since $6 = 3 \cdot 2$, we have two proper subgroups $\{0, 3\}$ and $\{0, 2, 4\}$. These can be written as the subgroup generated by 3 or $< 3 >$, and the subgroup generated by 2 or $< 2 >$. The coset representation generated by the subgroup $< 3 >$ is the collection of three cosets, $< 3 >= \{0, 3\}, < 3 > +1 = \{1, 4\}$, and $< 3 > +2 = \{2, 5\}$. The coset representation generated by the subgroup $\{2\}$ has only two cosets, $< 2 >= \{0, 2, 4\}$ and $< 2 > +1 = \{1, 3, 5\}$. Note that in both coset representations, each element of the group is represented exactly once.

3.2.7   Problems and Exercises:

1. On unit circles in the complex plane, approximately plot the 4'th, 5'th, 6'th, and 7'th roots of unity. Observe the motion of these roots of unity make when you multiply various ones by each other.

2. Prove that the roots of unity are a group under complex multiplication.

3. Prove Theorem 3.2.2.

4. The integers $\mathrm{mod}(N)$ are the set of positive integers $0, 1, 2, \cdot N - 1$. If the operation is addition, modulo(N) shows that this is a group. Namely, show closure, identity, and inverses. We refer to this group as $\mathbb{Z}_N$.

5. Show that the N'th roots of unity G_2 and $\mathbb{Z}_N$ are homomorphic under the map $w_N^k \to k$. Specifically, show that multiplication with the roots of unity is the same as addition with $\mathcal{Z}_N$.

6. For N even, show that the sum of the roots of unity is zero, or

$$\sum_{k=0}^{N-1} w_N^k = 0$$

without using a geometric sum argument. (Hint: Plot the roots on the unit circle and observe that there is a simple geometric, or arithmetic argument, which can easily be changed into a proof).

7. Are there any subgroups to the 7'th roots of unity? Are there any subgroups to the 6'th roots of unity? What are they? What are the conditions for there to be subgroups within the N'th roots of unity? How many proper subgroups will there be for arbitrary N?

3.3 The Fast Fourier Transform

One of the most important algorithms for fast computation is the Fast Fourier Transform or FFT. This was especially true when computing speed was very limited, computers were large and expensive, and problems were still very difficult. The obvious place to point to is the effort to get to the moon and back in the 1960s, with an onboard computer which was far less sophisticated than an average calculator today.

While the speed of computing has decreased the necessity of fast computation somewhat, problems just get larger. No matter how much computational ability you have, someone will have a problem with more variables than you are able to easily handle. An obvious example is hurricane prediction. To accurately predict the evolution of a hurricane, one needs temperature, humidity, wind speed, current direction, and perhaps other variables as thousands of locations. Thus, there is always a need for faster, more efficient computations.

The Fast Fourier Transform has been used in many forms, by a number of people over the years. The most commonly used version is attributed to Cooley and Tukey [6]. This is said to be the most referenced academic paper in history. While that claim may or may not be true, the algorithm and the subsequent uses have been central to a great number of scientific endeavors.

There are many papers and many books on this algorithm. We will look first at the basics and try to derive a simple version for two examples. There are people who have devoted careers to this algorithm and its following generalizations. Our goal is to understand some of the basics, in a short period of time.

3.3.1 Speed Enabling Algorithm

The primary achievement of the FFT is that the matrix vector product $F_N \vec{x}$ can be computed in $N \log_2(N)$ steps, rather than the usual N^2 steps which are required for simple matrix multiplication. This may not sound like much, but if $N = 1024$ which is rather simple, we are talking about the difference between approximately 1 million operations and 10 thousand operations, that is, a decrease in operations of a factor of 100.

If, however, we are working on video, or a two-dimensional picture, then this would be a decrease of $100^2 = 10000$. This implies that the FFT makes video compression 10000 times faster. To put this in perspective, every digital HDTV receiver from any cable or satellite company does some type of compression, so that you can get more channels and more movies over the same connection. These would be decoded 10000 times slower, without the FFT. This would mean that they would not be viable. Thus, if you watch a movie on one of the many movie channels you have at home tonight, take time to remember the FFT.

3.3.2 The simplest examples: F_2, F_4, and F_8.

The first Fourier matrix, F_2, is so simple as to be almost trivial. We will soon see that it is very significant, however, in the large scheme of things. The matrix is

$$F_2 = \frac{1}{\sqrt{2}} \begin{pmatrix} 1 & 1 \\ 1 & w_2 \end{pmatrix} = \frac{1}{\sqrt{2}} \begin{pmatrix} 1 & 1 \\ 1 & -1 \end{pmatrix}. \tag{3.7}$$

Obviously, this is an orthogonal matrix, and the prefactor $1/\sqrt{2}$ makes it orthonormal. Notice that there are two operations in computing a matrix-vector product, $F_2(\vec{g})$, which means that the number of operations is $2 \log_2 2$. One might think that these two operations are insignificant. We would argue, however, that they are extremely significant for the future and we denote them as the sum s_0 and difference d_0

$$s_0 = g_0 + g_1 \text{ and } d_0 = g_0 - g_1.$$

We will now move on to two other significant examples, where N=4 and N=8. We stick with powers of 2, since that is where things are nicest. We will find that things build on the most basic matrix F_2. We first examine the matrix F_4, and we separate the real and imaginary parts of the matrix for clarity. Thus, we have

$$F_4(\vec{g}) = \frac{1}{2} \begin{pmatrix} 1 & 1 & 1 & 1 \\ 1 & 0 & -1 & 0 \\ 1 & -1 & 1 & -1 \\ 1 & 0 & -1 & 0 \end{pmatrix} \begin{pmatrix} g_0 \\ g_1 \\ g_2 \\ g_3 \end{pmatrix} + \frac{i}{2} \begin{pmatrix} 0 & 0 & 0 & 0 \\ 0 & 1 & 0 & -1 \\ 0 & 0 & 0 & 0 \\ 0 & -1 & 0 & 1 \end{pmatrix} \begin{pmatrix} g_0 \\ g_1 \\ g_2 \\ g_3 \end{pmatrix}.$$

The goal is to multiply F_4 by $\vec{g} = (g_0, g_1, g_2, g_3)^t$ to get the Fourier Transform $(\hat{g}_0, \hat{g}_1, \hat{g}_2, \hat{g}_3)$ using only 8 additions and 8 multiplications, since $8 = 4 \log_2(4)$.

Simple observation of multiplying the above matrix by $(g_0, g_1, g_2, g_3)^t$, we see a lot of repetition. Namely, if we blindly multiplied the matrix by the vector, we would perform the same operations multiple times. To be specific, $g_0 + g_2$ occurs in 2 equations and $g_1 - g_3$ occurs in 2 equations, as do many other combinations. Given this, let us try to write these repeated combinations down in an orderly manner. Thus, we let

$$\begin{array}{ll} s_0 = \frac{1}{\sqrt{2}}(g_0 + g_2), & s_1 = \frac{1}{\sqrt{2}}(g_1 + g_3) \\ d_0 = \frac{1}{\sqrt{2}}(g_0 - g_2), & d_1 = \frac{1}{\sqrt{2}}(g_1 - g_3) \end{array}.$$

Notice that we are creating sums and differences of the even and odd components, respectively. In addition, the sum and difference operations are exactly the components of F_2. The reason for including the prefactor $1/\sqrt{2}$ will become apparent shortly. Now, if $\hat{g} = F_4 \vec{g}$, looking at the matrix above we can see that we have the following very simple result:

$$\begin{array}{ll} \hat{g}_0 = \frac{1}{\sqrt{2}}(s_0 + s_1), & \hat{g}_1 = \frac{1}{\sqrt{2}}(d_0 + id_1) \\ \hat{g}_2 = \frac{1}{\sqrt{2}}(s_0 - s_1), & \hat{g}_3 = \frac{1}{\sqrt{2}}(d_0 - id_1) \end{array}.$$

Note that we have used only 8 multiplications and additions, or calling an operation a multiplication and addition, 8 operations. Thus, we have the simple FFT for F_4.

Let us go back and investigate the above arithmetic, remembering F_2. If we examine our original sums and differences s_0, s_1, d_0, d_1, we see that they are given by

$$\begin{pmatrix} s_0 \\ d_0 \end{pmatrix} = F_2 \begin{pmatrix} g_0 \\ g_2 \end{pmatrix} \quad \text{and} \quad \begin{pmatrix} s_1 \\ d_1 \end{pmatrix} = F_2 \begin{pmatrix} g_1 \\ g_3 \end{pmatrix}. \tag{3.8}$$

Similarly, we have the final version given by

$$\begin{pmatrix} \hat{g}_0 \\ \hat{g}_2 \end{pmatrix} = F_2 \begin{pmatrix} s_0 \\ d_0 \end{pmatrix} \quad \text{and} \quad \begin{pmatrix} \hat{g}_1 \\ \hat{g}_3 \end{pmatrix} = F_2 \begin{pmatrix} d_0 \\ i * d_1 \end{pmatrix}. \tag{3.9}$$

At first glance, one would like to combine the above operations into two 2x2 matrices. Note, however, that the even and odd values of the final version swapped the variables s_1 and s_2 from the first operation. This can be written neatly as a combination of very sparse 4x4 matrices. We begin with

$$
\begin{pmatrix} s_0 \\ s_2 \\ s_1 \\ s_3 \end{pmatrix} = \begin{pmatrix} F_2 & 0 \\ 0 & F_2 \end{pmatrix} \begin{pmatrix} 1 & 0 & 0 & 0 \\ 0 & 0 & 1 & 0 \\ 0 & 1 & 0 & 0 \\ 0 & 0 & 0 & 1 \end{pmatrix} \begin{pmatrix} f_0 \\ f_1 \\ f_2 \\ f_3 \end{pmatrix} = M_2 M_1 f, \qquad (3.10)
$$

where it is understood that the 0's next to the F_2 matrices are 2x2 blocks of zeros.

Now, the final result is given by

$$
\begin{pmatrix} \hat{f}_0 \\ \hat{f}_1 \\ \hat{f}_2 \\ \hat{f}_3 \end{pmatrix} = \begin{pmatrix} F_2 & 0 \\ 0 & F_2 \end{pmatrix} \begin{pmatrix} 1 & 0 & 0 & 0 \\ 0 & 0 & 1 & 0 \\ 0 & 1 & 0 & 0 \\ 0 & 0 & 0 & i \end{pmatrix} \begin{pmatrix} s_0 \\ s_2 \\ s_1 \\ s_3 \end{pmatrix} = M_2 M_3 \vec{s}. \qquad (3.11)
$$

Therefore, we can write the entire FFT as

$$
\hat{f} = M_2 M_3 M_2 M_1 f.
$$

This simple example is the building block for the FFT. The FFT of order 2^q is built out of FFTs of order $2^{q-1}, 2^{q-2} \dots$.

Decomposing F_8

Now let us try to be a little more adventurous and find the FFT algorithm for F_8. Recall that we are allowed $8 \log_2(8) = 24$ operations, as opposed to the $N^2 = 64$ operations required for simple matrix multiplication. Let us once again consider the real and imaginary parts of the matrix. Since we know that F_8 has a $\frac{1}{\sqrt{8}}$ in front of it, we will delete this normalization factor for clarity. Thus, the real part of the matrix is given by

$$
\sqrt{8}Re(F_8) = \begin{pmatrix}
1 & 1 & 1 & 1 & 1 & 1 & 1 & 1 \\
1 & \frac{1}{\sqrt{2}} & 0 & -\frac{1}{\sqrt{2}} & -1 & -\frac{1}{\sqrt{2}} & 0 & \frac{1}{\sqrt{2}} \\
1 & 0 & -1 & 0 & 1 & 0 & -1 & 0 \\
1 & -\frac{1}{\sqrt{2}} & 0 & \frac{1}{\sqrt{2}} & -1 & \frac{1}{\sqrt{2}} & 0 & -\frac{1}{\sqrt{2}} \\
1 & -1 & 1 & -1 & 1 & -1 & 1 & -1 \\
1 & -\frac{1}{\sqrt{2}} & 0 & \frac{1}{\sqrt{2}} & -1 & \frac{1}{\sqrt{2}} & 0 & -\frac{1}{\sqrt{2}} \\
1 & 0 & -1 & 0 & 1 & 0 & -1 & 0 \\
1 & \frac{1}{\sqrt{2}} & 0 & -\frac{1}{\sqrt{2}} & -1 & -\frac{1}{\sqrt{2}} & 0 & \frac{1}{\sqrt{2}}
\end{pmatrix}.
$$

Similarly, we have for the complex part of the matrix,

$$
\sqrt{8}Im(F_8) =
\begin{pmatrix}
0 & 0 & 0 & 0 & 0 & 0 & 0 & 0 \\
0 & \frac{1}{\sqrt{2}} & 1 & \frac{1}{\sqrt{2}} & 0 & -\frac{1}{\sqrt{2}} & -1 & -\frac{1}{\sqrt{2}} \\
0 & 1 & 0 & -1 & 0 & 1 & 0 & -1 \\
0 & \frac{1}{\sqrt{2}} & -1 & \frac{1}{\sqrt{2}} & 0 & -\frac{1}{\sqrt{2}} & 1 & -\frac{1}{\sqrt{2}} \\
0 & 0 & 0 & 0 & 0 & 0 & 0 & 0 \\
0 & -\frac{1}{\sqrt{2}} & 1 & -\frac{1}{\sqrt{2}} & 0 & \frac{1}{\sqrt{2}} & -1 & \frac{1}{\sqrt{2}} \\
0 & -1 & 0 & -1 & 0 & -1 & 0 & 1 \\
0 & -\frac{1}{\sqrt{2}} & -1 & -\frac{1}{\sqrt{2}} & 0 & \frac{1}{\sqrt{2}} & 1 & \frac{1}{\sqrt{2}}
\end{pmatrix}.
$$

Now, we will try to arrange intermediate sums and differences as we did with F_4. To this end, we have

$$
\begin{aligned}
s_0 &= f_0 + f_2, & s_1 &= f_1 + f_3, & s_2 &= f_4 + f_6, & s_3 &= f_5 + f_7, \\
d_0 &= f_0 - f_2, & d_1 &= f_1 - f_3, & d_2 &= f_4 - f_6, & d_3 &= f_5 - f_7.
\end{aligned}
$$

We have now used 8 operations. Returning to the real portion the matrix, we see that the sum $s_4 = d_0 + d_4$ occurs twice, and doesn't seem to come from the above sums. Similarly, $d_4 = f_0 - f_4$ occurs in the 2'nd and 8'th lines and does not seem to be able to be represented by the above differences. By far, the dominant term in the real portion of the matrix is $d_5 = (d_1 - d_3)/\sqrt{2}$. We have used 3 more operations on s_4, d_4, and d_5 for a total of 11 operations.

We can now write the real portion of the Fourier Transform or

$$
Re
\begin{pmatrix}
\hat{f}_0 \\
\hat{f}_1 \\
\hat{f}_2 \\
\hat{f}_3 \\
\hat{f}_4 \\
\hat{f}_5 \\
\hat{f}_6 \\
\hat{f}_7
\end{pmatrix}
=
\frac{1}{\sqrt{8}}
\begin{pmatrix}
s_0 + s_1 + s_2 + s_3 \\
d_4 + d_5 \\
s_5 \\
d_4 - d_5 \\
s_0 - s_1 + s_2 - s_3 \\
d_4 - d_5 \\
s_5 \\
d_4 + d_5
\end{pmatrix}
$$

Counting operations, we have 3 in the first line, 1 in the second, 1 in the fourth, and 3 in the fifth. The rest of the vector has already been calculated, so we have a total of 11+8=19 operations for the real portion of the matrix.

Now for the imaginary portion of the matrix, we have a dominant term of $d_6 = (s_1 - s_3)/\sqrt{2}$. In addition, there is the term $d_7 = f_2 - f_6$ which occurs 4 times. We also need to term $d_8 = d_1 + d_3$, which occurs in the third and seventh lines. Thus, we require 3 more operations for a total of 22.

The result for the complex portion of the matrix is then summarized as

$$
Im \begin{pmatrix} \hat{f}_0 \\ \hat{f}_1 \\ \hat{f}_2 \\ \hat{f}_3 \\ \hat{f}_4 \\ \hat{f}_5 \\ \hat{f}_6 \\ \hat{f}_7 \end{pmatrix} = \frac{1}{\sqrt{8}} \begin{pmatrix} 0 \\ d_6 + d_7 \\ d_8 \\ d_6 - d_7 \\ 0 \\ -d_6 + d_7 \\ -d_8 \\ -d_6 - d_7 \end{pmatrix}.
$$

Now we have 4 more operations, in the 2'nd, 4'th, 6'th, and 8'th rows. Thus, we have 26 operations which leaves us 2 over budget.

Now, if we consider the real portion of the computation, we have $s_5 = s_0 + s_2$ and $s_6 = s_1 + s_3$ used twice. We can add these two operations and replace the real portion of the matrix with the reduced sums to get

$$
Re \begin{pmatrix} \hat{f}_0 \\ \hat{f}_1 \\ \hat{f}_2 \\ \hat{f}_3 \\ \hat{f}_4 \\ \hat{f}_5 \\ \hat{f}_6 \\ \hat{f}_7 \end{pmatrix} = \frac{1}{\sqrt{8}} \begin{pmatrix} s_5 + s_6 \\ d_5 + d_6 \\ s_5 \\ d_5 - d_6 \\ s_5 - s_6 \\ d_5 - d_6 \\ s_5 \\ d_5 + d_6 \end{pmatrix}.
$$

We eliminated 6 operations and replaced them with 4, so we are now at 24. We have succeeded.

3.3.3   The FFT in the Classic Case $N = 2^q$

We looked at the simple examples $F_2, F_4,$ and F_8 above. We were able to accomplish the goal of using $N \log_2(N)$ operations by precomputing the sums and differences of even and odd entries, individually. We would like to begin a more systematic approach for larger N

We will consider the classic case, where $N = 2^q$. Let us, therefore, start out with $N = 2N_1$ and examine the multiplication of each individual row. We move the $\sqrt{N}$ to the left side of our equation for simplicity.

We are going to switch the order of summation, so that we first sum over the even elements and then over the odd elements. We then have

$$\sqrt{N}\hat{f}(k) = \sum_{n=0}^{N-1} f(n)w_N^{kn}$$

$$= \sum_{j_2=0}^{1} \sum_{j_1=0}^{N_1-1} f(2j_1 + j_2)w_N^{k(2j_1+j_2)}$$

$$= \sum_{j_1=0}^{N_1-1} f(2j_1)w_N^{2kj_1} + f(2j_1 + 1)w_N^{k(2j_1+1)}$$

$$= \sum_{j_1=0}^{N_1-1} f(2j_1)w_N^{2kj_1} + \sum_{j_1=0}^{N_1-1} f(2j_1 + 1)w_N^{k(2j_1+1)}. \qquad (3.12)$$

We now recall from our study of the roots of unity that $w_N^2 = w_{N/2} = w_{N_1}$. Thus, we can rewrite 3.12 as

$$\sqrt{N}\hat{f}(k) = \sum_{j_1=0}^{N_1-1} f(2j_1)w_{N_1}^{kj_1} + w_N^k \sum_{j_1=0}^{N_1-1} f(2j_1 + 1)w_{N_1}^{kj_1}. \qquad (3.13)$$

We see from above that we have split the Fourier Transform of order N into two Fourier Transforms of order $N/2 = N_1$. One operates over the even and one over the odd indices. This is partially true, but remember that $k = 0, 1, 2, N - 1$, rather than stopping at N_1 if this was a Fourier Transform of order N_1. We must remember that the Fourier Transform is periodic, however, so the values above $k = N_1 - 1$ are repeats. The only value which changes is w_N^k. Thus, we can write

$$\hat{f}(k) = F_N(f) = \frac{1}{\sqrt{N}} \sum_{n=0}^{N-1} f(n)w_N^{kn}$$

$$= \frac{1}{\sqrt{2}} \frac{1}{\sqrt{N_1}} \left(\sum_{j_1=0}^{N_1-1} f(2j_1)w_{N_1}^{kj_1} + w_N^k \sum_{j_1=0}^{N_1-1} f(2j_1 + 1)w_{N_1}^{kj_1} \right)$$

$$= \frac{1}{\sqrt{2}} \left(F_{N_1}(f(2j_1))(k) + w_N^k F_{N_1}(f(2j_1 + 1))(k) \right)$$

$$= \frac{1}{\sqrt{2}} \left(F_{N_1}(f_e)(k) + w_N^k F_{N_1}(f_o)(k) \right) \qquad (3.14)$$

where we have designated the even and odd components of $f(n)$ by f_e and f_o. Note above that both of the Fourier Transforms $F_{N_1}(f_e)(k)$ and $F_{N_1}(f_o)(k)$ are N_1 periodic and that $k = 0, 1 \ldots N - 1$, so they both repeat once.

FFT order of complexity

We will now define $\eta(N)$ to be the upper bound on the number of operations necessary to calculate the Fourier Transform F_N. We have already shown that we can achieve the desired results $N \log_2(N)$ for $N = 2, 4$, and 8, so we will proceed by induction. Thus, we assume that for $\eta(N/2) = \eta(N/2) \log_2(N/2)$ and we must prove that the result holds for N.

Considering the above decomposition 3.14, we see that multiplication by w_N^k in the second term, and addition to the first, takes N operations. Since we are computing a Fourier Transform of length $N_1 = N/2$ twice, for the even and odd coefficients above, we have

$$\eta(N) = 2\eta(N/2) + N.$$

We have already shown that upper bounds for $\eta(2) = 2\log_2(2) = 2$ and that $\psi(4) = 4\log_2(4) = 8$, so it is proper to move to the induction step. Assuming that this is true for $N_1 = N/2$, we have from above that

$$\begin{aligned}
\eta(N) &= 2(N/2\log_2(N/2)) + N = N(\log_2(N/2) + 1) \\
&= N(\log_2(N/2) + \log_2(2)) = N(\log_2((N/2)2)) \\
&= N\log_2(N). \tag{3.15}
\end{aligned}$$

To understand the above, remember that $\log(ab) = \log(a) + \log(b)$. We have used this identity in reverse above, where $\log_2(N/2) + \log_2(2) = \log_2((N/2)2) = \log_2(N)$. Thus, we have

Theorem 3.3.1 (FFT Complexity) *An upper bound on the number of operations necessary to compute the Discrete Fourier Transform F_N if $N = 2^q$ with an FFT algorithm is $N\log_2(N)$.*

Revisiting F_4

We will now reexamine the structure of F_4 in terms of equation 3.14. Equation 3.14 states that for each $k = 0, 1, 2, 3$,

$$F_4(f)(k) = \frac{1}{\sqrt{2}} F_2(f_e)(k) + w_4^k F_2(f_o)(k). \tag{3.16}$$

Recall that the Fourier Transform is periodic, so $F_2(f_e)(k)$ will repeat. Thus, we can rewrite equation 3.16 in matrix form as

$$F_4(f) = \frac{1}{\sqrt{2}} \left(\begin{bmatrix} F_2 \\ F_2 \end{bmatrix} \begin{pmatrix} f_0 \\ f_2 \end{pmatrix} + \begin{pmatrix} 1 \\ i \\ -1 \\ -i \end{pmatrix} \cdot \begin{bmatrix} F_2 \\ F_2 \end{bmatrix} \begin{pmatrix} f_1 \\ f_3 \end{pmatrix} \right), \tag{3.17}$$

where the multiplication of the last two vectors is understood to be point by point multiplication of as in 3.14. Reverting to our earlier notation from equations 3.8 in the decomposition of F_4, this is equal to

$$F_4(f) = \frac{1}{\sqrt{2}} \left(\begin{pmatrix} s_0 \\ d_0 \\ s_0 \\ d_0 \end{pmatrix} + \begin{pmatrix} 1 \\ i \\ -1 \\ -i \end{pmatrix} \cdot \begin{pmatrix} s_1 \\ d_1 \\ s_1 \\ d_1 \end{pmatrix} \right) = \frac{1}{\sqrt{2}} \begin{pmatrix} s_0 + s_1 \\ d_0 + id_1 \\ s_0 - s_1 \\ d_0 - id_1 \end{pmatrix}. \qquad (3.18)$$

This is exactly the equations we obtained earlier. The difference is that we have a much more robust formalism to extend this into a full blown algorithm.

Revisiting F_8

Before we extend this discussion to a full algorithm, let us take the additional step of examining the FFT for F_8, in terms of equation 3.14. Equation 3.14 states that for each $k = 0, 1, 2, 3, 4, 5, 6, 7$,

$$F_8(f)(k) = \frac{1}{\sqrt{2}} F_4(f_e)(k) + w_4^k F_4(f_o)(k), \qquad (3.19)$$

where we have split the initial vector f into the subvectors $f_e = [f_0, f_2, f_4, f_6]$ and $f_o = [f_1, f_3, f_5, f_7]$. Thus, we can rewrite equation 3.19 in matrix form as

$$F_8(f) = \frac{1}{\sqrt{2}} \left(\begin{bmatrix} F_4 \\ F_4 \end{bmatrix} \begin{pmatrix} f_0 \\ f_2 \\ f_4 \\ f_6 \end{pmatrix} + \begin{pmatrix} 1 \\ w_8 \\ i \\ w_8^3 \\ -1 \\ w_8^5 \\ -i \\ w_8^7 \end{pmatrix} \cdot \begin{bmatrix} F_4 \\ F_4 \end{bmatrix} \begin{pmatrix} f_1 \\ f_3 \\ f_5 \\ f_7 \end{pmatrix} \right), \qquad (3.20)$$

where the multiplication of the last two vectors is understood to be point by point multiplication of as in 3.14.

We discussed the FFT for F_4 above, and we will now have to implement this algorithm to compute the four occurrences of F_4 in 3.19 above. We must first split the vector $f_e = [f_0, f_2, f_4, f_6]$ into the vectors $[f_0, f_4]$ and $[f_2, f_6]$ and then $f_o = [f_1, f_3, f_5, f_7]$ into $[f_1, f_5]$ and $[f_3, f_7]$. This is equivalent to taking the even and odd portions of our even and odd vectors, repeatedly.

Realize that this is merely the coset representation of $\mathbb{Z}_N$ generated by the subgroup $[0, 4]$. We refer to these as the cosets $< 4 > +k$, where $k = 0, 1, 2, 3$. We now utilize the decomposition of F_4 by computing F_2 on $f(< 4 >) = [f_0, f_4]$ and $f(< 4 > +2) = [f_2, f_4]$ and adding them according to 3.17 to get the left-hand vector in 3.20. We then multiply F_2 times the coset-generated vectors $f(< 4 > +1) = [f_1, f_3]$, and $f(< 4 > +3) = [f_3, f_7]$, and the correction factors to get the right-hand vector in 3.20. Adding these together gives us the final FFT for F_8.

Complete Decomposition of F_N, where $N = 2^q$.

We saw above that to compute F_8 efficiently, we needed to first compute F_2 on all of the cosets $f(< 4 > +k)$, where $k = 0, 1, 2, 3$. The cosets $< 4 >= [0, 4]$ and $< 4 > +2 = [2, 6]$ were then used to compute $F_4(f_e)$. Note, however, that $f_e = f(< 2 >)$, or the vector f evaluated at the subgroup or coset elements $< 2 >= [0, 2, 4, 6]$, and f_e is the vector at the coset elements $< 2 > +1 = [1, 3, 5, 7]$.

We introduced the general FFT with a single backward reduction 3.14 which we iterated on. We can now describe the complete FFT algorithm in a forward method using this approach. Note that the final decomposition of the vector f with 8 elements above was split into the vector on cosets with two elements. These were then individually multiplied by F_2, which recursively allows the computation of F_4 and then F_8. This is exactly how then general FFT is computed.

Coset Representation of Vectors for F_8

$$f(\mathbb{Z}_N) \qquad f(< 2 > +k) \qquad f(< 4 > +k)$$

$$
\begin{pmatrix} f_0 \\ f_1 \\ f_2 \\ f_3 \\ f_4 \\ f_5 \\ f_6 \\ f_7 \end{pmatrix}
\leftrightarrow
\begin{pmatrix} f_0 \\ f_2 \\ f_4 \\ f_6 \end{pmatrix} \begin{pmatrix} f_1 \\ f_3 \\ f_5 \\ f_7 \end{pmatrix}
\leftrightarrow
\begin{pmatrix} f_0 \\ f_4 \end{pmatrix} \begin{pmatrix} f_2 \\ f_6 \end{pmatrix} \begin{pmatrix} f_1 \\ f_5 \end{pmatrix} \begin{pmatrix} f_3 \\ f_7 \end{pmatrix}
. \qquad (3.21)
$$

Figure 3.3: We illustrate the process of breaking the original vector down into even and odd vectors, and then the final coset representation vectors for F_8 above.

We start with the cosets $< N/2 > +k$, where $k = 0, 1, 2, \ldots N/2 - 1$. We then compute the multiplication of F_2 times the $N/2$ generated subvectors $f(< N/2 > +k)$. These are then used to compute F_4 on the $N/4$ cosets $< N/4 > +k$, where $k = 0, 1, 2, \ldots N/4 - 1$, which corresponded to f_e and f_o above. Should N be 16, we would have had to do this one more time. In terms of the vectors and how they are back-computed it is helpful to look at sequence of coset vectors displayed in Figure 3.3.

The algorithm can be summarized as follows:

Forward Computation of the FFT Algorithm for general $N = 2^q$.

1. Compute F_2 on all of the subvectors $f(< N/2 > +k)$ where $k = 0, 1, ...$ $N/2 - 1$.

2. Use the above results and 3.14 to compute F_4 on all of the subvectors $f(< N/4 > +k)$, for $k = 0, 1, 2, ...N/4 - 1$, where N = 16.

3. Use the above results and 3.14 to compute F_8 on all of the subvectors $f(< N/8 > +k)$, for $k = 0, 1, 2, ..N/8 - 1$ where N = 32.

4. Iterate until you have computed F_N.

The most important thing to notice about the above algorithm is that there is only one set of matrix multiplication, i.e., those involving F_2 and the subvectors $f(< N/2 > +k)$ for $k = 0, 1, 2, ...N/2 - 1$. Thus, traditional matrix multiplication is almost completely eliminated from the algorithm. This is the significant factor in reducing the complexity from N^2, which is the typical matrix multiplication complexity, to $N \log_2(N)$, which is much more efficient.

3.3.4   The FFT when $N = N_1 N_2$

Oftentimes in computation, we choose $N = 2^q$ simply because the above algorithm is so fast and simple. Sometimes, however, we are given problems where N is not a power of two. We don't want to have to completely give up on fast computations. It turns out that we don't, and things are still much faster.

We will consider the case where N is arbitrary. Since N is an integer, we know that there is a prime factorization $N = p_1 p_2 p_3 \ldots p_k$, where all of the p_j are primes. The goal is to decompose the factorization down to the prime factors. Following as with the case $N = 2^q$ where we considered $N = 2N_1$, we will simply consider two factors $N = N_1 N_2$. From this, we should be able to get the general decomposition.

Recall our discussion of coset representations in the group $\mathbb{Z}_N$ 3.2.9. We want to consider an arbitrary integer $k \in \mathbb{Z}_N$, where $N = N_1 N_2$. We can then write it as either $k = j_1 N_2 + j_2$, or $k = j_2 N_1 + j_1$, where $j_1 = 0, 1, \ldots N_1 - 1$ and $j_2 = 0, 1, \ldots N_2 - 1$. In the case we considered where $N = 2^q = 2N_1$, we had obviously set $N_2 = 2$, which resulted in the even and odd decompositions. Since the coset representations are unique, we can decompose the sums in much the same way.

Decomposing the summation:

We now want to consider using this representation to break down the Fourier matrix multiplication or sum. We follow Benedetto's work fairly closely [1]. Recall that the Discrete Fourier Transform is given by

$$\hat{f}(k) = \frac{1}{\sqrt{N}} \sum_{n=0}^{N-1} f(n) w_N^{kn}.$$

Now if N factors into N_1 and N_2, then for any $n \in \mathcal{Z}_N$ we have $n = j_1 N_2 + j_2$, where $(j_2, j_1) \in \mathcal{Z}_{N_1} x \mathcal{Z}_{N_2}$. This embedding is one to one, so we can rewrite the sum as a double sum

$$\hat{f}(k) = \frac{1}{\sqrt{N}} \sum_{j_1=0}^{N_2-1} \sum_{j_2=0}^{N_1-1} f(j_1 N_2 + j_2) w_N^{k(j_1 N_2 + j_2)} \tag{3.22}$$

Now, we want to express each k as $k = k_2 N_1 + k_1$. Then, we have

$$\hat{f}(k) = \frac{1}{\sqrt{N}} \sum_{j_1=0}^{N_1-1} \sum_{j_2=0}^{N_2-1} f(j_1 N_2 + j_2) w_N^{k(j_1 N_2 + j_2)}$$

$$= \frac{1}{\sqrt{N}} \sum_{j_1=0}^{N_1-1} \sum_{j_2=0}^{N_2-1} f(j_1 N_2 + j_2) w_N^{(k_2 N_1 + k_1)(j_1 N_2 + j_2)}$$

$$= \frac{1}{\sqrt{N}} \sum_{j_1=0}^{N_1-1} \sum_{j_2=0}^{N_2-1} f(j_1 N_2 + j_2) w_N^{k_2 j_1 N_1 N_2} w_N^{k_2 N_1 j_2} w_N^{k_1 j_1 N_2} w_N^{k_1 j_2} \tag{3.23}$$

We realize that $w_N^{N_1 N_2} = w_N^N = 1$, and that $w_N^{N_1} = w_{N_2}$ and $w_N^{N_2} = w_{N_1}$. Thus, we have

$$\hat{f}(k) = \frac{1}{\sqrt{N}} \sum_{j_1=0}^{N_1-1} \sum_{j_2=0}^{N_2-1} f(j_1 N_2 + j_2) w_{N_2}^{k_2 j_2} w_{N_1}^{k_1 j_1} w_N^{k_1 j_2} \tag{3.24}$$

Now, let us change the order of summation and we get

$$\hat{f}(k) = \frac{1}{\sqrt{N}} \sum_{j_2=0}^{N_2-1} \left(\sum_{j_1=0}^{N_1-1} f(j_1 N_1 + j_2) w_{N_1}^{k_1 j_1} \right) w_{N_2}^{k_2 j_2} w_N^{k_1 j_2} \tag{3.25}$$

We will now concentrate on the interior sum for fixed k_1, k_2, and j_2. The first term, $j_1 = 0$, is the constant $f(j_2)$ regardless of the other values of k_1 and k_2.

Thus, we have to add multiply and add $N_1 - 1$ values to this, so we have used $(N_1 - 1)$ operations. We reorganize and get

$$\hat{f}(k) = \frac{1}{\sqrt{N_2}} \sum_{j_2=0}^{N_2-1} \left(\frac{1}{\sqrt{N_1}} \sum_{j_1=0}^{N_1-1} f(j_1 N_1 + j_2) w_{N_1}^{k_1 j_1} w_N^{k_1 j_2} \right) w_{N_2}^{k_2 j_2} \qquad (3.26)$$

and this multiplication implies that we have used N_1 operations to get the interior sum. Note that we have already summed over j_1 so the interior sum no longer depends on j_1. Similarly, it is independent of k_2. Therefore, we can refer to this sum as $S(j_2, k_1)$. The outer sum can be completed in N_2 operations, so it has taken $N_1 + N_2$ operations to complete this task, for fixed (k_1, k_2). Since there are N of them, this amounts to $N(N_1 + N_2)$.

If we have $N = 100^2 = 10,000$, then we have reduced the number of multiplications from $N^2 = 100^4 = 10^8$ or 100 million to $10,000(100 + 100) = 2,000,000$. This is obviously a factor of 50, which is good. This amounts to a $2N\sqrt{N}$ algorithm. We would like to do better.

The obvious story is to complete the decomposition by decomposing sum into the prime factorization $N = p_1 p_2 p_3 ... p_k$. This can be done recursively as we did with the case $N = 2^q$.

3.3.5   Problems and Exercises:

1. Develop a Fast algorithm for evaluating the FFT on 6 points or decomposing F_6. Utilize some of the techniques used in the examples for F_4 and F_8. The straightforward calculation takes 36 operations. How many can you reduce this to?

2. Decompose the matrix F_8 into a product of F_4 matrices using equation 3.14. Then, decompose this further into a product of F_2 matrices, using the approach of 3.16.

3. Compare equation 3.14 to the direct method which we used in equations 3.10 and 3.11 to decompose F_4. Show that they are the same.

4. Use equation 3.14 to decompose F_8 making sure it is correct.

5. **Challenging:** Write the FFT in a compact way for F_8. Compare it to a standard DFT. Can you do it for F_{16}?

6. **Challenging Project:** Write the FFT code for an arbitrary number 2^k, or for F_{2^k}?

3.4   Discrete Convolution and Correlation

We first introduced the idea of convolution on $L^2[a,b]$ in Chapter 2. We will now introduce it in the setting of the Discrete Fourier Transform. One of the great advantages and uses for the FFT is the fast computation of discrete convolutions. Instead of directly computing the convolution, we use the FFT to compute the Fourier Transforms, and as before, we multiply them pointwise.

Let us now clarify the idea of discrete convolution and some of the differences and similarities to convolution as we studied it in Chapter 2. We begin with two vectors $g, h \in R^n$. We denote these by $g = \{g_m\}_{k=0}^{N-1}$ and $h = \{h_n\}_{n=0}^{N-1}$. Let us consider their Discrete Fourier Transforms, denoted by $\hat{g} = Fg$ and $\hat{h} = Fh$, where F is the discrete Fourier matrix.

Now if $\hat{g}_k$ is the kth entry in the vector $\hat{g}$, it follows that $\hat{g}_k = 1/\sqrt{N} \sum_{m=0}^{N-1} g_m w^{km}$ and $\hat{h}_k = 1/\sqrt{N} \sum_{n=0}^{N-1} h_n w^{kn}$, so their pointwise product is

$$\hat{g}_k \hat{h}_k = \frac{1}{N} \left(\sum_{m=0}^{N-1} g_m w^{km} \right) \left(\sum_{n=0}^{N-1} h_n w^{kn} \right)$$

$$= \frac{1}{N} \left(\sum_{m=0}^{N-1} \sum_{n=0}^{N-1} g_m h_n w^{k(m+n)} \right) \qquad (3.27)$$

We want to reorganize this sum into a single Fourier Series by letting $j = m + n$, which means that $n = j - m$, and we get

$$\hat{g}_k \hat{h}_k = \sum_{j=0}^{N-1} \left(\sum_{m=0}^{N-1} g_m h_{j-m} \right) w^{kj} = \sum_{j=0}^{N-1} a_j w^{kj}.$$

Thus, the product of the Fourier coefficients of g and h is the Discrete Fourier Transform of the vector of coefficients

$$\{a_j\}_{j=0}^{N-1} = \{ \sum_{m=0}^{N-1} g_m h_{j-m} \}_{j=0}^{N-1}. \qquad (3.28)$$

Before moving forward, we must comment on the notation. Namely, we have referred to the terms h_{j-m} in the above formula for the coefficients a_j. This is to be interpreted in "modulo" arithmetic. Namely, $h_{j-m} = h_{j-m+N}$ whenever $j - m$ is negative. To see that this makes sense, let us consider one concrete example with $N = 4$. In this case, we would have

$$
\begin{aligned}
\hat{g}_k \hat{h}_k &= (g_0 w^0 + g_1 w^k + g_2 w^{2k} + g_3 w^{3k})(h_0 w^0 + h_1 w^k + h_2 w^{2k} + h_3 w^{3k}) \\
&= (g_0 h_0 + g_1 h_3 + g_2 h_2 + g_3 h_1) w^0 \\
&\quad + (g_0 h_1 + g_1 h_0 + g_2 h_3 + g_3 h_2) w^k \\
&\quad + (g_0 h_2 + g_1 h_1 + g_2 h_0 + g_3 h_3) w^{2k} \\
&\quad + (g_0 h_3 + g_1 h_2 + g_2 h_1 + g_3 h_0) w^{3k}.
\end{aligned}
\tag{3.29}
$$

Note above that, for instance, we have $(g_1 w^1)(h_3 w^3) = g_1 h_3 w^4 = g_1 h_3 w^0$, since $w_0 = w_4 = 1$. Similarly, $(g_2 w^2)(h_3 w^3) = g_2 h_3 w^5 = g_2 h_3 w^4 w^1 = g_2 h_3 w^1$. Thus, the sum of the two coefficients for w^k is always $k mod(N)$. Tracing the progression of the subscripts of h above, in the sum 3.28 it might appear that for $j = 0$ the subscripts of h are $0, -1, -2, -3$, but with modulo arithmetic they correspond to the simple multiplication 3.29 which are $0, 3, 2, 1$.

Note that the coefficients operations in (3.29) can be organized in the matrix equation

$$
Hg = \begin{pmatrix} h_0 & h_3 & h_2 & h_1 \\ h_1 & h_0 & h_3 & h_2 \\ h_2 & h_1 & h_0 & h_3 \\ h_3 & h_2 & h_1 & h_0 \end{pmatrix} \begin{pmatrix} g_0 \\ g_1 \\ g_2 \\ g_3 \end{pmatrix}.
\tag{3.30}
$$

Thus, we now define the discrete convolution of two discrete sequences.

Definition 3.4.1 (Discrete Convolution) *Given two vectors in $\mathbb{C}^N$, $g = \{g_k\}_0^{N-1}$ and $h = \{h_k\}_0^{N-1}$ we define the discrete convolution of g and h to be*

$$
g * h(j) = \sum_{m=0}^{N-1} g_m h_{j-m},
\tag{3.31}
$$

for $j = 0 \ldots N - 1$.

Let us also define some notation.

Definition 3.4.2 (Pointwise Multiplication) *The pointwise multiplication of two vectors $g = \{g_k\}$ and $h = \{h_k\}$ is denoted and defined by $g.h = \{g_k h_k\}$.*

We have already discovered or proven the major theorem of this section as was stated in (3.28). We now state it for later use. This is a fundamental theorem of Fourier Analysis.

Theorem 3.4.1 (Discrete Convolution Theorem) *Let $g = \{g_k\}_0^{N-1}$ and $h = \{h_k\}_0^{N-1}$ be two vectors in R^N. Then the convolution $g * h$ has the Discrete Fourier Transform*

$$
F(g * h) = F(g).F(h) = \left\{ \hat{g}(k) \hat{h}(k) \right\}_{k=0}^{N-1}
$$

Written more specifically, the Fourier Transform of the convolution is equivalent to pointwise multiplication of the individual Fourier vectors.

Diagonalization is equivalent to changing matrix multiplication into pointwise multiplications. Thus, the Fourier Transform diagonalizes convolution. We must add another definition and corollary.

Definition 3.4.3 (Discrete Correlation) *Given two vectors in R^N, $g = \{g_k\}_0^{N-1}$ and $h = \{h_k\}_0^{N-1}$ we define the discrete correlation of g and h to be*

$$g \star h(j) = \sum_{m=0}^{N-1} g_m h_{m-j}, \tag{3.32}$$

for $j = 0 \dots N - 1$.

Note that the only difference between convolution and correlation is that the vectors are reversed. Thus, if a convolution/correlation is a symmetric function, there is no difference.

Corollary 3.4.1 *Let $g = \{g_k\}_0^{N-1}$ and $h = \{h_k\}_0^{N-1}$ be two vectors in R^N. Then the correlation $g \star h$ has the Discrete Fourier Transform*

$$F(g \star h) = F(g).\overline{F(h)} = \left\{ \hat{g}(k)\overline{\hat{h}(k)} \right\}_{k=0}^{N-1}$$

Written more specifically, the Fourier Transform of the correlation of two vectors, is equivalent to pointwise multiplication of the first, times the conjugate of the second.

The proof is identical to that for convolution and is left as an exercise.

3.4.1   Circulant Toeplitz Matrices

Let us now return to the matrix representation of convolution which we introduced in (3.30),

$$Hg = \begin{pmatrix} h_0 & h_3 & h_2 & h_1 \\ h_1 & h_0 & h_3 & h_2 \\ h_2 & h_1 & h_0 & h_3 \\ h_3 & h_2 & h_1 & h_0 \end{pmatrix} \begin{pmatrix} g_0 \\ g_1 \\ g_2 \\ g_3 \end{pmatrix}. \tag{3.33}$$

Note that the entries along the diagonal, as well as the super- and subdiagonals, are all the same. This is called a Toeplitz matrix. You can define this by the fact that the matrix entries $a_{i,j} = a_{i-j}$, or they do not depend independently on i and j, but only on $i - j$.

The above matrix has an additional quality. The first row is simply circulated or translated with modulo arithmetic, so that $ai, j = a_{1,j-i}$. This comes directly from the definition of convolution. A general circulant Toeplitz matrix is of the form

$$T = \begin{pmatrix} t_0 & t_1 & t_2 & t_3 & \cdots & t_N \\ t_N & t_0 & t_1 & t_2 & \cdots & t_{N-1} \\ t_{N-1} & t_N & t_0 & t_1 & \cdots & t_{N-2} \\ & \cdot & \cdot & \cdot & \cdot & \cdot \\ t_1 & t_2 & t_3 & t_4 & \cdots & t_0 \end{pmatrix}. \tag{3.34}$$

The interesting thing about circulant Toeplitz matrices is that they are all diagonalized by the Fourier Transform.

Theorem 3.4.2 (Circulant Toeplitz Diagonalization) *Let T be any circulant Toeplitz matrix. Then*

$$T = F_N^{-1} D F_N. \tag{3.35}$$

where F_N is the Fourier matrix.

This is simply a matrix restatement of Theorem (3.4.1). The unresolved matrix in (3.35) is the diagonal D. The entries on the diagonal D are the Fourier coefficients of either a convolution, or a correlation. This just depends on the ordering of the coefficients. We illustrate a convolution in (3.33). The correlation matrix is given by

$$Hg = \begin{pmatrix} h_0 & h_1 & h_2 & h_3 \\ h_3 & h_0 & h_1 & h_2 \\ h_2 & h_3 & h_0 & h_1 \\ h_1 & h_2 & h_3 & h_0 \end{pmatrix} \begin{pmatrix} g_0 \\ g_1 \\ g_2 \\ g_3 \end{pmatrix}. \tag{3.36}$$

In the case of the correlation matrix, the top row h of the correlation matrix is simply the top row of the Toeplitz matrix (3.34), t, or $h = t$. The diagonal entries in D are then just the conjugates of the Fourier Transform of the first row t.

In the case of a convolution matrix, the mapping is slightly more involved. Let us illustrate this by starting with an arbitrary circulant Toeplitz such as (3.34). Identifying this with the convolution matrix (3.33), we see that the coefficient $h_0 = t_0$. Things change then, $h_1 = t_N, h_2 = t_{N-1}, h_3 = t_{N-2}, h_4 = t_{N-3}$, or in general $h_k = t_{N-k+1}$, when $k > 0$. More simply stated, the convolution vector would be

$$h = [t_0, t_N, t_{N-1}, t_{N-2}...t_1]. \tag{3.37}$$

The coefficients of the diagonal D in the matrix multiplication Tg can then be computed as $\hat{h}_k$, using the Discrete convolution theorem. Similarly, the whole operation Tg can be computed using the Fourier Transform and the products $\hat{h}_k\hat{g}_k$, followed by the Inverse Fourier Transform.

We will use this analysis on a number of occasions in the following applications.

3.5 Applications of Convolution and Correlation

Convolution and correlation are fundamental to Fourier Analysis and to Physics. This is one of the reasons why the Fourier Transform is such an important tool to science. Instead of computing Tg directly which would require N^2 operations, we will use (3.35), which requires two Fourier Transforms and the multiplication by the diagonal. The final operation count is therefore $2Nlog_2(N) + N = N(2log_2(N) + 1)$. Obviously, for N large, $2log_2(N) + 1$ is much smaller than N. What does convolution mean. Let us examine discrete convolution equation 3.32

$$g * h(j) = \sum_{m=0}^{N-1} g_m h_{j-m}, \qquad (3.38)$$

for $j = 0 \ldots N - 1$ and the result of Theorem 3.4.1, namely

$$F(g * h) = F(g).F(h) = \left\{ \hat{g}(k)\hat{h}(k) \right\}_{k=0}^{N-1}$$

through a number of basic examples.

Realize that there is no difference between the Fourier Transforms of convolution and correlation except for the conjugate on the Fourier Transform of g.

3.5.1 Derivatives via Convolution

One very common use of convolution or correlation is to compute derivatives. We will present several possible ways to do this, although there are many possible ways.

Approximate Derivatives

To begin with, we will begin with a very simple function, namely $g(t) = \sin(t)$ for the 32 discrete points $t_k = [0 : 511]/512 * 2 * \pi$. Now mathematically, we always define

$$g'(t) = \lim_{h \to 0} \frac{g(t+h) - g(t)}{h}. \tag{3.39}$$

Since we only have discrete data in this section, we can only consider an approximate derivative. One candidate for this might be

$$g'(t_n) \approx \frac{g(t_{n+1}) - g(t_n)}{h_n}. \tag{3.40}$$

This is called a forward difference. The term $h_n = (2\pi)/512$ is the distance between subsequent points, which is consistent with the definition of the derivative.

Returning to the idea of the circulant Toeplitz matrix, the forward difference matrix would look like

$$D = 1/h_n \begin{pmatrix} -1 & 1 & 0 & 0 & 0 & \cdots\cdots & 0 \\ 0 & -1 & 1 & 0 & 0 & \cdots & 0 \\ 0 & 0 & -1 & 1 & 0 & \cdots & 0 \\ 0 & 0 & 0 & -1 & 1 & \cdots & 0 \\ . & . & . & . & . & . & . \\ 1 & 0 & 0 & 0 & 0 & \cdots & -1 \end{pmatrix}. \tag{3.41}$$

Returning to equation (3.37) to find the convolution vector, we see that it is $h = 1/h_n * [-1, 0, 0, ...0, 1]$.

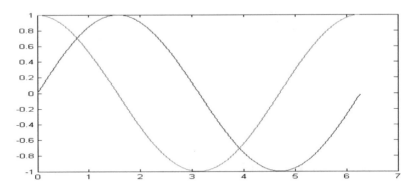

Figure 3.4:   We illustrate a discrete realization of a sine above, and the discrete derivative calculated via the above convolution approximation 3.40. It is nearly exact to the actual derivative, a cosine.

We now calculate the Fourier Transform of both g and h and then the Inverse Fourier Transform of $\hat{g}\hat{h}$ which is the convolution. This is illustrated in Figure 3.4. The approximate derivative in the above calculation and plot was accurate to .0067. This is a very nice result, but the function was simple, and the data exact.

Noisy Data

Oftentimes, we have data that is not exact. Instead, we have the correct data plus some random noise such as described in Section 6.5.1. We will now revisit the last problem, with the same function but with a reasonable amount of noise added to the system. Repeating Figure 3.4, but with noise. We see the result in Figure 3.5. The derivative is nowhere near the derivative illustrated above. This is due to noise. Derivatives are very sensitive to small changes. The derivative is linear, so the derivative shown is the derivative of the function, plus the derivative of the noise. The derivative of the noise does not exist in a true sense and is swamping the derivative of the noise.

Formal or Exact Derivatives

Using the matrix representation (3.5), and noting that the conjugate of the Fourier matrix is its inverse, we can represent any vector in the form

$$g_n = \frac{1}{\sqrt{N}} \sum_{k=-N/2}^{N/2-1} \hat{g}_k \overline{w}^{kn}.$$

In a continuous representation, t would replace n. Furthermore, $w^{kn} = e^{ikn} = \cos(kn) + i\sin(kn)$. Thus thinking of n as a continuous time variable, we should

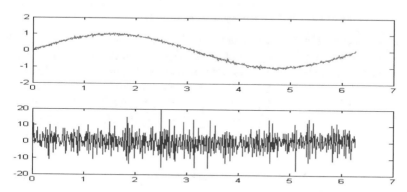

Figure 3.5: We illustrate a discrete realization of a sine plus a small amount of random noise above. The discrete derivative calculated via the above convolution approximation 3.40. Note that the derivative is nothing like the derivative we would anticipate.

consider the formal derivative (differentiating with respect to n), and therefore, we get

$$\frac{d}{dn}\overline{w}^{kn} = \frac{d}{dn}e^{-ikn} = (-ik)w^{kn}$$

which results in the "formal derivative"

$$g'_n = \frac{1}{\sqrt{N}} \sum_{k=-N/2}^{N/2-1} (-ik\hat{g}_k)\overline{w}^{kn}. \tag{3.42}$$

The question would be, "Is this a true derivative?" The answer is an emphatic yes, if you are thinking of a finite continuous sum of sines and cosines. The answer is exact. The error in the derivative calculated with the standard approximation above was .0067. The error with this method was 4e-14, which is essentially machine precision for MATLAB. In other words, this is exact.

This should not be a surprise. To understand, realize that if you have $t = [0 : N - 1]/N * 2 * \pi$, then $\cos(kt)$ is real part of the vector $\{w^{kn}\}_{n=0}^{N-1}$. Similarly, $\cos(kt)$ is the real part of $\{w^{(N-k)n}\}_{n=0}^{N-1}$. Since the imaginary part of both $\sin(kt)$ is orthogonal, the result will be the same from the dot products of $\cos(kt)$ against to two conjugate columns of F. Thus, the cosine coefficients of the Discrete Fourier Transform are symmetric in that sense. The sine coefficients of the discrete Fourier Transform will be antisymmetric, since $\{w^{kn}\}_{n=0}^{N-1} = \{\overline{w}^{(N-k)n}\}_{n=0}^{N-1}$, or $w^k = \overline{w}^{N-k}$. In addition, the sine coefficients are imaginary. Thus, you have real cosine coefficients, which are symmetric, and imaginary sine coefficients, which are antisymmetric.

To understand how multiplying by ik times the coefficients makes sines into cosines, it exactly transforms imaginary antisymmetric coefficients into real symmetric coefficients. Thus, it is an exact way to calculate derivatives.

1. Prove the Corollary 3.4.1 to Theorem 3.4.1.

2. **Forward Difference:** Compute the forward difference of $\sin(2 * t)$, where $t = [0 : 31]/32 * 2 * pi$ (t on 32 points from 0 to 2π) in 3 ways: first, using the Toeplitz matrix as in (3.41). Use the MATLAB command $>> D = \text{toeplitz}(c,r)$, where c is the first column and r is the first row of D for 32 points; second, using the discrete convolution theorem; third, using the discrete correlation theorem. All three of these computations should be identical to machine accuracy or approximately 10^{-15}.

3. **Backward Difference:** The backward difference approximation for a derivative is given by

$$f(x_n) \approx \frac{f(x_n) - f(x_{n-1}))}{dx}.$$

As in Problem 2, construct the Toeplitz matrix for this approximate derivative, and compute it in three ways: a) by matrix multiplication, b) by the discrete convolution theorem, and c) by the discrete correlation theorem. All three of these computations should be identical to machine accuracy, or approximately 10^{-15}.

4. **Symmetric Difference:** The symmetric difference approximation for a derivative is given by

$$f(x_n) \approx \frac{f(x_{n+1}) - f(x_n - 1))}{2dx}.$$

As in Problem 2, construct the Toeplitz matrix for this approximate derivative, and compute it in three ways: a) by matrix multiplication, b) by the discrete convolution theorem, and c) by the discrete correlation theorem. All three of these computations should be identical to machine accuracy, or approximately 10^{-15}.

5. Write self-contained programs which perform both of the above algorithms in files of the type filename.m (see MATLAB's explanation of functions). Use either the convolution or correlation approach. Test the programs on a variety of functions such as $\sin(kt)$ for various k values. Is there a difference in accuracy when k is large? Can you explain this?

6. **Formal or Exact Derivatives:** Write a self-contained program which computes the formal or exact derivative of a function, as defined in (3.42). Be careful to make sure that your multipliers ik match up correctly with the proper frequency coefficients. Test this on a few functions $\sin(kt)$ to assure accuracy.

3.5.2   Averaging

One way to suppress the noise is to attempt to average it away as explained in Section 6.5.2. Namely, let us try to take the average of the noise, over a finite interval. That would involve, for instance, the averaging vector $[1, 1, 1, \ldots, 1, 1, 1]/6$.

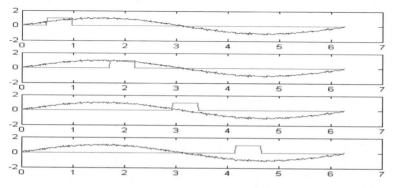

Figure 3.6:   We illustrate the process of averaging via convolution above. We have a function plus noise. The averaging filter, represented by the square wave, is moved sequentially along the function, and local averages of the function plus noise are calculated.

Let us first try to regain the original signal by trying to average it, or suppress the noise in a number of ways.

We begin by taking local averages of the function. This process, via convolution, is illustrated in Figure 3.6.

We used a variety of averaging filters, beginning with one that was 6 pixels wide, and then using one 20 pixels wide. The results are shown in Figure 3.7.

Averaging and Taking Derivatives

We have shown in Figure 3.7 that we can reduce the amount of noise by averaging the values of the function or vector. The question remains, "Can we still take the derivative of the function, after averaging away the noise?" We have illustrated taking the derivative, as above with the standard two-step difference after averaging in Figure 3.8.

Notice that the derivative is accurately represented if we average by a factor of 20. This was done by using the discrete difference formula 3.40. Another question is whether or not the formal derivative, represented in equation 3.42, also yields similar results. The answer is no. The reason is that multiplying every Fourier coefficient by ik results in seriously amplifying the noise. Thus, the result is not at all useful.

3.5.3   Narrowband and Nonlinear filtering

Narrowband filtering

Let us return to our example of a sinusoid with noise, and consider the absolute value of its Fourier Transform. This is shown in Figure 3.9.

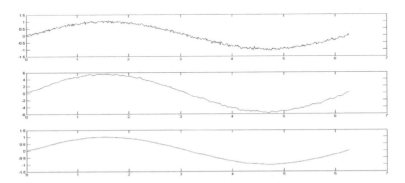

Figure 3.7: We illustrate the averaging of a sine plus noise above. The top graph is the sine plus noise. The second graph illustrates a local average with 10 local points used for the average. The third graph has 20 points used for averaging. Note that the noise is approximately suppressed with a 20 point average.

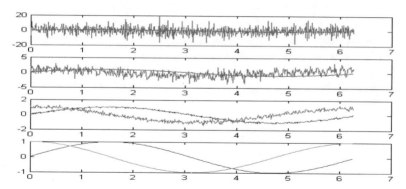

Figure 3.8:   We illustrate a discrete realization of a sine plus a small amount of noise above. The derivative is then calculated via the approximation 3.40. Note that the first derivative is nothing like the derivative we would anticipate. If we average the sine by factors of 6, and 20, then we see in the third frame that we get something resembling the derivative of the sine, i.e., nearly a cosine. The desired result is shown for comparison in the fourth graph at the bottom.

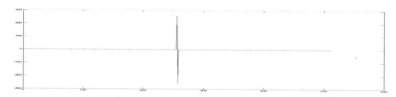

Figure 3.9:   The imaginary part of the Discrete Fourier Transform of the noisy sinusoid is shown above. This is shifted, to have zero frequency in the center. Notice the ±1 values around the origin, indicating $\sin(t)$. The noise is at a lower level but at every frequency.

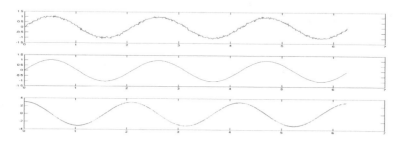

Figure 3.10:   The first figure above shows a sinusoid with noise. In the second, the out-of-band noise is suppressed, and only the frequencies which are of interest are kept above. Notice that nearly all of the out-of-band noise is suppressed. The error is only .004, which is much lower than the noisy result. In the third graph, we took the formal or exact derivative of the filtered result and compared it to the actual derivative $(3\cos(3t))$ without noise. Once again, the answer was nearly exact, with an error of about .02.

The question might be, "if we know that we only have low-frequency signals, why do we keep the other coefficients?" Very good question, and the answer is often there is no reason. In Figure 3.10, we only keep the coefficients which we feel are relevant. One might ask why we could "anticipate" that those frequencies are the only frequencies which are relevant. Oftentimes in science and engineering, we know ahead of time what the signal looks like. For instance, in communications, we know what signals we are transmitting. Thus, we look for those signals, using a linear narrowband filter.

Nonlinear filtering

In nonlinear filtering, we separate the "good," and "bad" frequencies according to some rule, which is nonlinear. For instance, we choose to only keep those frequency components which are above a certain threshold. These are the significant ones, and the others are arbitrarily chosen to be noise, since noise is generally represented by many small coefficients. We will use multiple different frequencies for this illustration, namely the function $f(t) = \sin(3 * t) + \sin(15 * t) + \sin(30 * t)$.

For our algorithm, we find the mean of the absolute value of the Fourier coefficients. We then retain any coefficients which are larger than $1/10^{th}$ of the mean. This is nonlinear, because it operates dependent upon the relative magnitude of the coefficients. An narrowband filter, as above, would not necessarily get all of these coefficients.

The results of this nonlinear filtering and subsequent derivatives with and without filtering are shown in Figure 3.11. Note that not only is the function adequately recovered from noise, but also its derivative.

The question arises, When does nonlinear filtering not work? The answer is simple. We had several dominant frequencies in our answer. If the original signal had many frequencies, then we probably could not have used this simple technique quite so easily. The moral is the same in all situations, however. If you know something about the signal, then you can suppress noise. If you do not know anything, then you have a difficult problem.

3.5.4 Matched Filtering

One of the most useful applications for the DFT or FFT is the matched filter. This is used extensively in radar, communications, remote imaging, and many other applications. We will revisit these later, but will give an introduction to the idea here.

The basic idea is that you are looking for one type of signal $s(t)$. In communications, you are generally trying to listen for a signal which was transmitted from a distant substation. You know its form, and frequency, or you know the signal $s(t)$. What you are interested in is the time at which it is transmitted. Once again, this process is easy if there is no noise, but there always is. Furthermore, when distance is increased, the signal power will necessarily be decreased, and therefore, the signal power will be reduced compared to the noise.

Correlation is the natural tool to use for the matched filter. We assume that the receiver is "hearing" a signal

$$r(t) = s(t - T) + n(t)$$

where $n(t)$ is random noise. We then use correlation to calculate

$$r(t) \star s(t) = \int r(\tau)s(t - \tau)d\tau.$$

The reason why the matched filter works is simple. Consider the matched filter when the noise is not present. Then, we have

$$r(t) \star s(t) = \int s(\tau - T)s(t - \tau)d\tau.$$

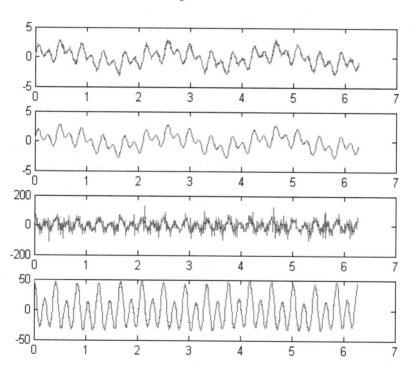

Figure 3.11: We begin with the function described above, and a version with substantial noise, with the difference being .16. We then show in the second plot the original with nonlinear filtering. The difference is then .02, and the graphs are much more similar. The third graph shows the formal derivative of the original function and its noisy derivative. The final graph shows the result of using nonlinear filtering before differentiation. The difference in this final result is approximately .03, while the noisy derivative had an error of more than 1 or was not close.

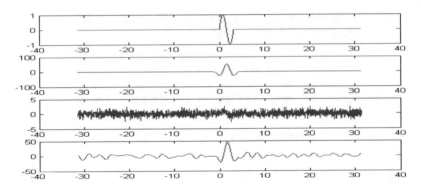

Figure 3.12: The first graph shows the signal, which is to be "matched", by the matched filter. The second graph shows what is often called the autocorrelation function. This is the correlation of the signal with itself. The third graph shows a realistic situation with a lot of noise, and the signal somewhat buried. The fourth graph shows the result of using the matched filter on the noisy signal in the third graph. The noise is suppressed and the location of the signal is clearly shown.

Notice that his is simply a series of shifted inner products. We know from the Cauchy–Schartz inequality that

$$\langle s(\tau - T)s(t - \tau)\rangle$$

is maximized only when $T = \tau$. Thus, if we are trying to identify the location and presence of $s(t - T)$, the best option is to look for the maximum value of the matched filter.

We illustrate this with a simple example. We will choose our signal to be

$$s(t) = \begin{cases} \sin(4t) & \text{if } 0 \le t \le 2\pi \\ 0 & \text{otherwise} \end{cases}$$

We will now consider the problem of detecting this function between -10π and 10π with and without the presence of serious noise. This is illustrated in Figure 3.12. The noisy signal can be detected with the matched filter.

3.5.5 Problems and Exercises:

1. Use MATLAB to add noise to $\sin(kt)$ for some $k > 1$ (the command is $\sin(k*t)+ c*\text{randn}(\text{size}(t))$; for some level of c which you can adjust, perhaps starting with c = .05). Try to take the derivative of this noisy function.

2. Repeat Problem 3 with the function

$$f = [zeros(1, 256), ones(1, 256), zeros(1, 128), 2 * ones(1, 256), zeros(1, 128)].$$

Do things seem to change? Can you take the derivative?

3. Take derivatives of the noisy functions in Problems 3 and 4 using both the exact and approximate derivative approximations. Which seems better?

4. Try to average the functions from Problems 3 and 4, and take their derivatives after averaging. Were you able to get a satisfactory result?

5. Use narrowband filtering before taking the derivatives from Problems 3 and 4. Were you able to get a satisfactory result?

6. Use nonlinear filtering to repeat the problem above.

3.6 Computing Sine and Cosine Expansions Using the FFT: Compression

In Chapter 2, we introduced the idea that we could use only the cosine transform, or only the sine transform, in order to represent a function. Let us now devise methods to utilize this idea cleanly with the FFT. We could try to derive direct algorithms for the sine or cosine transforms, but rather, we will try to utilize variants with the standard FFT.

One application of the cosine transformation is for compression. Namely, given a function, or a vector, try to find a minimal number of coefficients which will accurately represent that function. Thus, you are compressing the size of a data file. We have shown in Chapter 2 that this is a viable possibility with the Cosine transform. We will now do this in the discrete setting.

We will assume throughout that we have a vector f_k, where $k = 0 \ldots N - 1$. We will not assume anything about this vector, except that it is real. If it is complex, the real and imaginary parts can be dealt with separately. We will embed this vector into a larger vector and then find its respective Cosine or Sine expansion.

3.6.1 Cosine expansions

The difficult part of doing this is the discrete nature of the transform. It is easy to make something "odd," or "even," with a continuous function, but with discrete functions, it is not so obvious.

Let us begin by considering the basic "cosine", which is the real part of w_n^k, where $k = 0 \cdots N - 1$, or $\cos(2\pi k/N)$. Let us think now about expanding this beyond the interval, rather to $k = -N \ldots N - 1$. Then, we do have a truly even function since $\cos(2\pi k/N) = \cos(-2\pi k/N)$, or if we are going to extend a function f_k, where $k = 0 \ldots N - 1$ to $k = -N \ldots N - 1$, we need to have $f_k = f_{-k}$. This brings up the problem of what to do with f_{-N}. It turns out it does not matter much.

Let us use the notation of MATLAB, where f_k, $k = 0 \cdots N - 1$, can be written as $f(1 : N)$, since the subscript 0 does not exist. If we simply use the equation

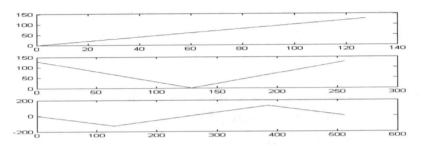

Figure 3.13: We illustrate our first simple vector, and the even and odd extensions which we constructed. While both the even and odd extensions contained more points, their FFTs allowed us to store the original vector with less information.

$newf = [f(N), f(N : -1 : 2), f(1 : N)]$, then this will be an "odd" function, for the purposes of the FFT. Namely, the terms will be cosine terms, and therefore, the FFT will be purely real, up to machine precision. One point which is arbitrary here is why we started with $F(N)$. It turns out that this is somewhat arbitrary and may be replaced with other values. By making it $f(N)$, the continuity of the function at the endpoints is preserved.

Example

Let us do a concrete example. Suppose that we want to find a compact FFT for the vector $0, 1, 2, 3 \ldots 127$. Because of the discontinuity at the endpoint, transitioning from 127 back to 0 (recall that the FFT is still periodic in k), this will not be efficiently represented by the FFT. If, however, we consider the function outlined above as $127, 127, 126, 125, 124, 123, \ldots 2, 1, 0, 1, 2, \ldots 126$, 127, then note that there is no discontinuity at the endpoints.

Taking the FFT of this function, we would like to use the isometry to see how many coefficients are necessary to represent this vector with an relative accuracy of 1%. For this measure of accuracy, if we call the Discrete Fourier Transform $\hat{f}_k$, we want

$$\frac{\sqrt{\sum_{k=0}^{N-1} \hat{f}_k^2 - \sum_{k=0}^{M-1} \hat{f}_k^2}}{\sqrt{\sum_{k=0}^{N-1} \hat{f}_k^2}} < .01, \tag{3.43}$$

where M is the smallest number such that this happens. Note that since the negative and positive cosine coefficients are the same, we only have to check these from 0 to $N-1$, even though our FFT will have $2N$ coefficients.

Doing this with the vector $0, 1, 2, 3 \ldots 127$, we find that only 9 coefficients are needed for this accuracy. From the viewpoint of compression, we have used only 7% of the original number of coefficients. Let us assume that we try to make this an odd function; namely, we use instead $-128, -127, -126, \cdots -1, 0, 1, \ldots 125, 126, 127$, which is the odd extension of the original vector. Taking the FFT, we find that this is truly a sine expansion, or all of the nonzeros coefficients are complex. Using the same metric, it would take 103 coefficients to have the same accuracy. Thus, there is very little compression or only $103/128$.

3.6.2 Sine expansions

Sine expansions can be used similar to Cosine expansions, if the first and last elements of the vector are essentially zero. This is because, the first and last elements of any Sine function are zero. Let us revisit the above example with a similar goal in mind. We begin with the vector $0, 1, 2, 3, \ldots 127$. We will now extend it to make the first and last elements zero, or consider $0, 1, 2, 3, \ldots, 127, 128, 127, 126, \ldots 2, 1$. Now, we can make it an odd function, by extending it to

$$0, -1, -2, -3, \cdots -127, -128, -127, -126, -125, \qquad (3.44)$$
$$\ldots -3, -2, -1, 0, 1, 2, 3, \ldots 127, 128, 127, 126, \ldots 2, 1$$

An FFT of this will give us a sine series.

Note that we have gone from a vector of length 128 to one of length 512. If we use the same measure, we find that the first 12 coefficients are sufficient to represent the function with the same accuracy. Further observation of the coefficients shows that every other sine coefficient is 0, however, so we only need 6 coefficients. As a result, this is more efficient than the cosine expansion. By organizing into a larger computational setting, we are able to retain far fewer coefficients. This is one of the goals of compression.

3.6.3 Piecewise Cosine and Sine expansions

The above discussions focused primarily on removing the *induced discontinuities* from a signal. By induced discontinuities, we mean the discontinuities which are created when the beginning and end of the signal are not the same.

These become induced discontinuities due to the fact that the Fourier representation tries to make them periodic, and since the beginning and end are not similar, there is an induce discontinuity.

Often, there are *natural discontinuities* in a signal or function. Thus, there is a break in the signal at some point in time. We know that the Fourier Transform does not represent discontinuities well, so we must do something to reduce the effects of these breaks.

One obvious, quick, and relatively useful technique is to represent only a limited portions of the signal a cosine or sine expansion. Thus, if we have a vector which has 256 coefficients, we represent every 32 coefficients with its own Cosine or Sine transform. In this case, we would be utilizing 8 different Cosine representations and one original function. The hope is that *natural discontinuities* will not disable the compression capabilities of the transform for the whole signal, but only for that portion where the discontinuity exists.

This is illustrated in Figure 3.14.

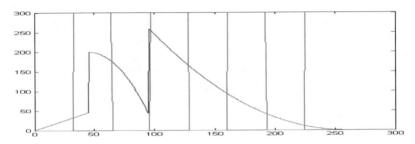

Figure 3.14: We illustrate the idea of a piecewise Cosine expansion above. The signal is split into 8 signals, as illustrated by the green lines. Each individual signal is then compressed with a Cosine expansion. The discontinuities are then isolated to a two of the individual expansions.

3.6.4 Problems and Exercises:

1. (a) Write a program which will take a signal and compute its Cosine transform, utilizing the methods described above. (b) Write a program which will decide how many significant coefficients are necessary from the cosine transform for a given input tolerance (.01, or .001 for instance)— in other words, the minimum number of coefficients for a desired accuracy, as defined in (3.43).

2. (a) Write a self-contained program which will take a signal and compute its Sine transform, utilizing the methods above. (b) Once again, write program which will decide how many coefficients are necessary for a given input tolerance, as defined in (3.43).

3. Compare the Sine and Cosine transforms above on a number of example. First, see whether they give the same results as in the example above. Then, test them on various other examples. Examine the final transformations to see whether all of the coefficients are nonzero.

4. Write a program which will do a piecewise Cosine transform, as illustrated in Figure 3.14. Have the program decide how many coefficients are necessary for a given tolerance. Test this on a piecewise continuous function such as 3.14. Is this more efficient than the standard transform?

3.7 Chapter Project

1. **Symmetric Difference:** The symmetric difference approximation for a derivative is given by

$$f(x_n) \approx \frac{f(x_{n-1}) - f(x_n))}{2dx}.$$

As in Problem 2, construct the Toeplitz matrix for this approximate derivative, and compute it in three ways: a) by the matrix multiplication, b) by the discrete convolution theorem, and c) by the discrete correlation theorem. All three of these computations should be identical to machine accuracy or approximately 10^{-15}.

2. **Formal or Exact Derivatives:** Write a self-contained program which computes the formal or exact derivative of a function, as defined in (3.42). Be careful to make sure that your multipliers ik match up correctly with the proper frequency coefficients. Test this on a few functions such as $\sin(kt)$ to assure accuracy. Once again, this should be exact to machine precision.

3. Use MATLAB to add noise to $\sin(kt)$ for some $k > 1$ (the command is >> fnoisy $= \sin(k*t)+ c*$randn(size(t)); for some level of c which you can adjust, perhaps starting with c $=$ .05). 1) Try to take the derivative of this noisy function, using both of the above derivative algorithms. Which works best? 2) Try to use averaging, narrowband filtering, or nonlinear filtering to reduce the noise before taking the derivative. Which works best?

4. Take the derivative of the function

$$f = [zeros(1, 256), ones(1, 256), zeros(1, 128), 2 * ones(1, 256), zeros(1, 128)],$$

with both the approximate and exact derivatives. Are these what you expected?

5. **Noise Reduction** Create a test function which has zeros on 1024 points, except for a simple sinusoid with 3 cycles which is 32 data points long in the middle (i.e., $\sin([0:63]/64*2*pi*3)$. 1) Add noise at different levels to this test function, and use the matched filter to detect the sinusoid. At what noise levels does this seem possible? 2) Increase the length of the sinusoid to 64 data points with 6 cycles. Now, use the matched filter with various noise levels as before. Is this function easier to find? Can you guess why?

6. **Cosine Transform** Write a self-contained program which will take a signal and compute its Cosine transform, utilizing the methods described above. Write a program decide how many significant coefficients are necessary for a given input tolerance (.01, or .001 for instance). Use (3.43) to determine the error. Make sure that the FFT is only returning real coefficients, with the exception of small machine-level errors.

7. **Sine Transform** Write a self-contained program which will take a signal and compute its Sine transform, utilizing the methods above. Once again, write another program to decide how many coefficients are necessary for a given input tolerance using (3.43) as the measure of error.

8. Compare the Sine and Cosine transforms above on a number of examples, such as t and $t.^2$. First see whether they give the same results as in the example above. Then, test them on various other examples. Examine the final transformations to see how many coefficients are significant at different error levels.

9. Develop a fast algorithm for evaluating the FFT on 6 points or decomposing F_6. Utilize some of the techniques used in the examples of F_4 and F_6 in this book. The straightforward calculation takes 36 operations. How many can you reduce this to?

10. [**Bonus Project:**] Using the iterative methods of Chapter 3, write an FFT code for 64 points.

Chapter 4
The Fourier Transform

4.1 From Fourier Series to the Fourier Transform

The transition from Fourier Series which were introduced in Chapter 2, to the Fourier Transform is similar to the transition from the dot product to the inner product. When dealing with functions, the question became "How many samples of the function are sufficient to approximate, accurately, the dot product that would be used to compare vectors?". The answer is that the safe way to make sure you have enough samples is to not sample, but rather use the integral, or inner product, instead of the dot product. That way convergence issues are not an issue, and everything was certain, while preserving the geometry of the dot product.

The similar question arises here. We have Fourier Series for any function on any finite interval. What happens if we look at a function which does not reside on a finite interval $[-T, T]$? How do we deal with it? We would still like to consider its Fourier representation. One answer is to concentrate on functions which are extensions of $L^2[-T, T]$, and which we define through the following.

Definition 4.1.1 *We say that a function $f(t) : \mathbb{R} \to \mathbb{C}$, where $\mathbb{C}$ are the complex numbers, is in $L^2(\mathbb{R})$ if the square of its integral is finite, which defines*

$$L^2(\mathbb{R}) = \left\{ f \mid \int_{-\infty}^{\infty} |f(t)|^2 dt < \infty \right\}.$$

We want to note that we can extend or restrict functions from $L^2[-T, T]$ to $L^2(R)$, and let the following two definitions guide us.

Electronic supplementary material The online version of this chapter (https://doi.org/10.1007/978-1-4939-7393-4_4) contains supplementary material, which is available to authorized users.

T. Olson, *Applied Fourier Analysis*,
https://doi.org/10.1007/978-1-4939-7393-4_4

Definition 4.1.2 (Extension) *Note that any function in $L^2[T,T]$ represents by extension a function in $L^2(\mathbb{R})$. Specifically, let $f(t) \in L^2[a,b]$, and let*

$$f_e^T(t) = \begin{cases} f(t) & \text{if } t \in [-T,T] \\ 0 & \text{if } |t| > T. \end{cases}$$

We know from Theorem 2.3.3 that we can represent $f_e(t)$ in the Fourier Series

Similarly, if we have a function which is in $L^2(\mathbb{R})$ we want to consider its restriction to $[-T,T]$.

Definition 4.1.3 (Restriction) *Note that any function in $L^2(\mathbb{R})$ also represents a function in $L^2[-T,T]$ by restriction. Specifically, let $f(t) \in L^2(\mathbb{R})$, and let*

$$f_r^T(t) = \{f(t) \text{ if } t \in [-T,T].$$

We know from Theorem 2.3.3 that we can represent $f_e^T(t)$, and f_r^T in a Fourier Series, as given by

Theorem 4.1.1 *Let $f(t)$ be any function in $L^2[-T,T]$. Then we can represent $f(t)$ in a series as*

$$f(t) = \sum_{k=-\infty}^{\infty} c_k \frac{e^{i\pi kt/T}}{\sqrt{2T}} \tag{4.1}$$

where

$$c_k = \frac{1}{\sqrt{2T}} \int_{-T}^{T} f(t)e^{i\pi kt/T} dt. \tag{4.2}$$

The question which we are most interested in is "Do the Fourier representations of the restriction, $f_r^T(t)$, somehow converge to some type of Fourier representation for $f(t)$?" To be more specific, remember that if $f(t) \in L^2(\mathbb{R})$, then the functions $f_r^T(t)$ are approximations to the Fourier coefficients of the function $f(t)$. Mathematically we know that in $L^2(\mathbb{R})$, we have

$$\lim_{T\to\infty} \|f_r^T(t) - f(t)\|_2 = 0. \tag{4.3}$$

We would like to believe from 4.3 that since we are representing something coming infinitely close to $f(t) \in L^2(\mathbb{R})$, the representation must in the long run become a good representation of $f(t)$. This is absolutely not true. The representation 4.1 is T periodic, and as a result, it will repeat itself throughout. Thus, the representation 4.1 will be good in $[-T,T]$, and irrelevant outside of that region (*the Fourier Series will repeat the function from $[-T,T]$, on $[T,3T]$, $[3T,5T]$, etc.*).

Thus, somehow we must make a transition to a new kind of representation. Recall that at this time we are considering a countable number of coefficients

c_k, where k is an integer. Let us consider these coefficients instead, as a function of a continuous variable s.

Definition 4.1.4 (The Continuous Fourier Transform) *For any function $f(t) \in L^2(\mathbb{R})$, we define its continuous Fourier Transform to be*

$$\hat{f}(s) = \frac{1}{\sqrt{2\pi}} \int_{-\infty}^{\infty} f(t)e^{ist} dt. \tag{4.4}$$

Now, note that the coefficients 4.2 are samples of $\hat{f}(s)$, as long as we are dealing with a function which is zero outside of $[-T, T]$, or with the restriction f_r^T.

If $f(t) \in L^2(\mathbb{R})$, however, it is not immediately clear that 4.4 is even well defined. Specifically, if $s = 0$, there is no guarantee that the integral exists and is finite. To assist in the understanding and transfer of this process, we introduce the following

Definition 4.1.5 *We say that a function $f \in L^1(\mathbb{R})$ if $\int_{-\infty}^{\infty} |f(t)| dt < \infty$.*

Note that if f is bounded and $f \in L^1(\mathbb{R})$, then $f \in L^2(\mathbb{R})$. The reason that we look at $L^1(\mathbb{R})$ as an intermediary step is because we can definitely state the following

Theorem 4.1.2 *Let $f(t) \in L^1(\mathbb{R})$. Then for every s,*

$$\hat{f}(s) = \frac{1}{\sqrt{2\pi}} \int_{-\infty}^{\infty} f(t)e^{ist} dt$$

is bounded and exists.

Proof: Boundedness is easy, since

$$|\hat{f}(s)| \leq \frac{1}{\sqrt{2\pi}} \int_{-\infty}^{\infty} |f(t)e^{ist}| dt = \frac{1}{\sqrt{2\pi}} \int_{-\infty}^{\infty} |f(t)| dt < \infty.$$

To understand existence, remember that we say that $\int_{-\infty}^{\infty} g(t) dt$ exists if $\lim_{T \to \infty} \int_{-T}^{T} g(t) dt$ exists. Now since $\int_{-\infty}^{\infty} |f(t)| dt < \infty$, it follows that for every ϵ there is a $T(\epsilon)$ such that $\int_{-\infty}^{\infty} |f(t)| dt - \int_{-T(\epsilon)}^{T(\epsilon)} |f(t)| dt < \epsilon$, or that $\int_{t > T(\epsilon)} |f(t)| < \epsilon$. This implies that

$$\left| \int_{t > T(\epsilon)} f(t)e^{ist} dt \right| < \epsilon,$$

which shows that the $\lim_{T \to \infty} \int_{-T}^{T} f(t)e^{ist} dt$ exists. $\square$

Thus, it is easier to initially consider the Fourier Transform for functions in $L^1(\mathbb{R})$, although we will concentrate on the $L^2(\mathbb{R})$ situation.

Searching for an inverse transformation

We would like to not only have a Fourier Transform, which we have shown is well defined for $f \in L^1(\mathbb{R})$ above. We would also like to have an Inverse Fourier Transform, or a representation for $f(t)$ in terms of these coefficients, or of the Fourier Transform.

Returning to the Fourier Series expansion 4.1, the coefficients can be rewritten as

$$c_k = \sqrt{\frac{\pi}{T}} \hat{f}\left(\frac{\pi k}{T}\right),\tag{4.5}$$

and therefore the function can be represented on $[-T, T]$ as

$$f(t) = \sum_{k=-\infty}^{\infty} \sqrt{\frac{\pi}{T}} \hat{f}\left(\frac{\pi k}{T}\right) e^{-ikt}.\tag{4.6}$$

The Fourier Series 4.1 represents a function on $[-T, T]$ with the above coefficients, or samples of the continuous Fourier Transform. If we want to represent a function on a larger interval, say $[-2T, 2T]$, then we need the additional coefficients and the representation

$$f(t) = \sum_{k=-\infty}^{\infty} \sqrt{\frac{\pi}{2T}} \hat{f}\left(\frac{\pi k}{2T}\right) e^{-ikt/2}.$$

Thus, to represent a function on an interval which is twice the length, we need two times more coefficients, or twice as many samples of the continuous Fourier Transform. This argument works for an interval of arbitrary size. One might be tempted to think that this would allow one to eventually use the series 4.1 to represent $f(t)$ on the whole real line. Remember that the series 4.1 will always be T periodic, however.

All is not lost, however. Similarly, to switching from a dot product to an inner product, we can now switch from the sum in Theorem 4.1.1 to an integral, and the representation is then valid.

Theorem 4.1.3 (The Inverse Fourier Transform and Representation) *If $f(t) \in L^2(\mathbb{R})$, then we can represent $f(t)$ by its Fourier Transform, or*

$$f(t) = \frac{1}{\sqrt{2\pi}} \int_{-\infty}^{\infty} \hat{f}(s) e^{-ist} ds.\tag{4.7}$$

Before we can prove this theorem, we need to have a few more tools. After we have these, we will return to this theorem.

4.2   The Fourier Isometry from $L^2(\mathbb{R})$ to $L^2(\mathbb{R})$

Let us consider the relationship between $f(t)$ and its Fourier Transform $\hat{f}(s)$. Recall that we have an isometry between any function $f(t) \in L^2[-T, T]$ and l^2. We would like to extend this relationship to this more general setting. We will begin with functions $f(t) \in L^2(\mathbb{R})$ which are zero except in a finite interval $[-T_0, T_0]$. Now let $T > T_0$ be arbitrarily large. We have from the isometry on $L^2[-T, T]$ 2.3.4

$$\int_{-T}^{T} |f(t)|^2 dt = \sum_{k} |c_k^T|^2 \tag{4.8}$$

where c_k^T denotes the Fourier coefficient for $f(t)$ on $[-T, T]$. Recalling that $f(t)$ is zero outside of $[-T, T]$, we see that

$$\int_{-\infty}^{\infty} |f(t)|^2 dt = \int_{-T}^{T} |f(t)|^2 dt = \sum_{k} |c_k^T|^2. \tag{4.9}$$

We see that the left term is independent of T. Returning to the representation of c_k^T as a function of $\hat{f}(s)$ 4.5, we have for any $(T > T_0)$ that

$$\int_{-\infty}^{\infty} |f(t)|^2 dt = \sum_{k} |c_k^T|^2 = \sum_{k} \frac{\pi}{T} \left| \hat{f}\left(\frac{\pi k}{T}\right) \right|^2. \tag{4.10}$$

Now, if we define $s_k = \frac{\pi k}{T}$, then from Freshman Calculus $\delta_s = \frac{\pi}{T}$, and we have

$$\int_{-\infty}^{\infty} |f(t)|^2 dt = \sum_{k} |c_k^T|^2 = \sum_{k} |\hat{f}(s_k)|^2 \delta_s. \tag{4.11}$$

The left hand side is independent of T, while the right hand side does depend upon T, so we let $T \to \infty$, meaning that $\delta_t \to 0$, and thus, we have

$$\int_{-\infty}^{\infty} |f(t)|^2 dt = \int_{-\infty}^{\infty} |\hat{f}(s)|^2 ds. \tag{4.12}$$

Note that continuity of the right hand side assures the convergence of the integral in a Riemann sense. The functions $f(t) \in L^2(\mathbb{R})$ which are zero outside a finite interval are dense in $L^2(\mathbb{R})$, so the equality must hold for all functions in $L^2(\mathbb{R})$, and we have proven another important theorem.

Theorem 4.2.1 *Let $f(t) \in L^2(\mathbb{R})$, and let $\hat{f}(s)$ be its continuous Fourier Transform as defined in Definition 4.1.4. Then we have*

$$\int_{-\infty}^{\infty} |f(t)|^2 dt = \int_{-\infty}^{\infty} |\hat{f}(t)|^2 dt. \tag{4.13}$$

In addition, the Fourier Transform defined in Definition 4.1.4 is an isometry from $L^2(\mathbb{R})$ to $L^2(\mathbb{R})$.

Proof: We have only proven that distances are preserved. We have not proven that the inner products are preserved.

If we check, however,

$$\|f(t) \pm g(t)\|^2 = \|f(t)\|^2 + \|g(t)\|^2 \pm 2\langle f(t), g(t)\rangle$$

which implies that inner products must be preserved also. Detailed arguments of this are left as exercises.

4.2.1   Problems and Exercises:

1. Finish the proof of Theorem 4.2.1. Namely, show that if distances are preserved, then inner products must also be preserved.
2. If f is real, show that $\hat{f}(s)$ is conjugate symmetric, namely that $Re(\hat{f}(s))$ is even, or symmetric, and $Im(\hat{f}(s))$ is odd, or antisymmetric. This called conjugate symmetry because $\hat{f}(s) = \overline{\hat{f}(-s)}$.
3. Show that if f is even, $\hat{f}(s)$ is real and symmetric, and if f is odd, $\hat{f}(s)$ is purely imaginary and antisymmetric.
4. Calculate the Fourier Transform of $\chi_\pi(t)$.
5. Calculate the Fourier Transform of $\chi_\pi(t)\cos(t)$.
6. Calculate the Fourier Transform of $\chi_\pi(t)\cos^2(t)$.
7. Calculate the Fourier Transform of $\chi_\pi(t)t$.
8. Calculate the Fourier Transform of $\chi_\pi(t)t^2$.
9. **Challenging** Show directly that the Fourier isometry is preserved between $\chi_{[-\pi,\pi]}(t)$, and its Fourier Transform, which is $\sin(\pi t)/(\pi t)$. In other words, show directly that they have the same L^2 norm.

4.3   The Basic Theorems of Fourier Analysis

We want to examine some of the basic identities, or theorems associated with the Fourier Transform on $L^2(\mathbb{R})$. We have established that for every function $f(t) \in L^2(\mathbb{R})$ we have a function $\hat{f}(s) \in L^2(\mathbb{R})$ as its Fourier Transform. We need to understand, both mathematically and intellectually how these functions interact under basic operations. One of the fundamental properties of Fourier Analysis is that there are simple expressions for how basic mathematical operations affect the Fourier Transform.

4.3.1   Differentiation

A natural question is how does differentiation affect the Fourier Transform. The answer is very simple, and does a great deal to explain why we utilize the Fourier Transform in the beginning. The goal of mathematics is to simplify complex problems which are of importance to science and mathematics. Differential equations come up in the description of processes in nearly every scientific discipline. The Fourier Transform easily describes the existence and the form of derivatives.

Theorem 4.3.1 *Let $f(t) \in L^2(\mathbb{R})$ and let $\hat{f}(s) \in L^2(\mathbb{R})$ be its Fourier Transform. If $(-is)\hat{f}(s) \in L^2(\mathbb{R})$, then it follows that $f'(t)$ exists as an element of $L^2(\mathbb{R})$, and that its Fourier Transform is*

$$\mathcal{F}(f'(t)) = (-is)\hat{f}(s).$$

Proof: Let us begin with the Inverse Fourier Transform

$$f(t) = \frac{1}{\sqrt{2\pi}} \int \hat{f}(s)e^{-ist}ds.$$

Now we want to differentiate $f(t)$ or consider

$$f'(t) = \frac{d}{dt} \int \frac{1}{\sqrt{2\pi}}\hat{f}(s)e^{-ist}ds.$$

The simple conclusion to this theorem involves moving the derivative operator inside of the integral resulting in

$$f'(t) = \frac{1}{\sqrt{2\pi}} \int \frac{d}{dt}\hat{f}(s)e^{-ist}ds = \frac{1}{\sqrt{2\pi}} \int (-is)\hat{f}(s)e^{-ist}ds.$$

We now have that

$$f'(t) = \mathcal{F}^{-1}\left((-is)\hat{f}(s)\right),$$

which proves our theorem. The step which we haven't verified is whether or not we are allowed to change the order of integration and differentiation. The answer is that yes, you can change the order of integration and we refer to [6] for details of the arguments.

It is hard to overstate the consequences of Theorem 4.7.6. First, it allows one to reduce the differential equations to multiplication by polynomials in the transform domain. Second, it implies that if a function has derivatives, or multiple derivatives, then its Fourier Transform must decay quickly in the transform domain, since $(is)f(\hat{s})$ must be square integrable, or that

$$\int s^2 |\hat{f}(s)|^2 ds < \infty$$

or that $\hat{f}^2(s) = o(1/s^3)$ (Recall that $\int_1^\infty \frac{1}{s} ds = \infty$).

Iterating this argument yields a more general theorem

Theorem 4.3.2 *Let $f(t) \in L^2(\mathbb{R})$ and let $\hat{f}(s) \in L^2(\mathbb{R})$ be its Fourier Transform. If $(-is)^m \hat{f}(s) \in L^2(\mathbb{R})$, then it follows that the mth derivative of f, $f^m(t)$ exists as an element of $L^2(\mathbb{R})$, and that its Fourier Transform is*

$$\mathcal{F}(f'(t)) = (-is)^m \hat{f}(s).$$

Thus, smoothness in the time domain is identical to the rate of decay of the Fourier Transform. Let us pause at this time, however, to remind the user that the Fourier Transform is an isometry from $L^2(\mathbb{R}) \rightarrow L^2(\mathbb{R})$. Thus, all arguments can be reversed. If a function is smooth in frequency, then it must decay quickly in time. Similarly, if it is not smooth in frequency, then it cannot decay quickly in time.

4.3.2   Translation

We now definite the translation operator. Once again, the Fourier Transform does a very good job of providing a precise description of the action in the transform domain, when a function is translated in the other domain.

Definition 4.3.1 (Translation) *Let $f(t) \in L^2(\mathbb{R})$. We refer to translating a function $f(t)$ by a factor of a as the operator $T_a f(t) = f(t - a)$.*

The corresponding theorem is

Theorem 4.3.3 *If $f(t) \in L^2(\mathbb{R})$ and $\hat{f}(s)$ is its Fourier Transform then*

$$\mathcal{F}(T_a(f(t))) = \mathcal{F}(f(t - a)) = \hat{f}(s)e^{isa}. \qquad (4.14)$$

Thus, translation corresponds to multiplication by a exponential function in frequency.

Let us think about this for a second before we move on. Think of a Fourier Series of a function on an interval. If we translate that function, the same frequencies will still be present, just translated... suggesting that the Fourier Transform shouldn't change. The Theorem above verifies this, in that the magnitude of the Fourier coefficients, $\hat{f}(s)$ is not changed. The thing that changes is its phase because of the multiplication by e^{isa}. This is similar to the phase change between a $\cos(kt)$ and $\sin(kt)$, i.e., one is identical to the other, but off by one half a period. Thus, translation changes sines into cosines, and vice versa. Thus, this theorem states that *translation does not change the magnitude of the frequency components, or Fourier Transform, but rather induces a phase change on individual coefficients, or collectively the entire Fourier Transform.*

Proof: We proceed methodically as in the former proof.

$$\mathcal{F}(f(t-a)) = \frac{1}{\sqrt{2\pi}} \int f(t-a)e^{ist}dt.$$

We now make the change of variables on the right, $x = t - a$, and we get

$$\mathcal{F}(f(t-a)) = \frac{1}{\sqrt{2\pi}} \int f(s)e^{is(x+a)}dx = e^{isa}\frac{1}{\sqrt{2\pi}} \int f(s)e^{isx}dx = e^{isa}\hat{f}(s).$$

The consequences of the properties of the Fourier Transform presented in this chapter will become very evident in the next chapter, and other subsequent applications chapters. While there may seem to be a lot to understand, a basic understanding of these principles will take a mathematician/scientist/engineer a very long way.

4.3.3   Convolution and Correlation

We now define a fundamental concept, which is central to a great deal of applications. Convolution and correlation are central to radar processing, digital communications, and many other applications. The basic idea is a very simple one. Given a signal or image which may be degraded, you may want to use local averages to reduce noise, or suppress clutter. Thus, the moving average is a key to a great deal of signal processing. We now define convolution and correlation and prove the necessary theorems.

Definition 4.3.2 (Convolution) *If $f(t)$ and $g(t)$ are functions in $L^2(\mathbb{R})$, then we define the convolution of f and g to be*

$$f * g(t) = \frac{1}{\sqrt{2\pi}} \int f(\tau)g(t-\tau)d\tau.$$

Thus, the convolution of g against f is a moving average, which is centered at time t.

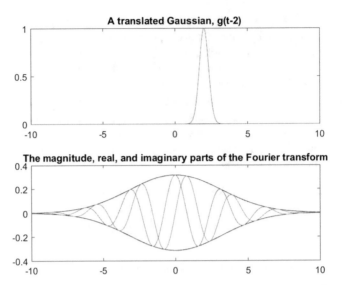

Figure 4.1: In the top figure, we plot a translated Gaussian $g(t) = e^{-(t-2)^2}$ in blue. At the bottom, we plot the absolute value of its Fourier Transform, as well as the real part of its Fourier Transform in blue, and the complex part of its Fourier Transform in red.

The question should be asked, "Why do we use $g(\tau - t)$ instead of $g(t - \tau)$?". This is a good question and leads to another definition.

Definition 4.3.3 (Correlation) *If $f(t)$ and $g(t)$ are functions in $L^2(\mathbb{R})$, then we define the correlation of f and g to be*

$$f \star g(t) = \frac{1}{\sqrt{2\pi}} \int f(\tau)g(\tau - t)d\tau.$$

What is the difference between convolution and correlation? There essentially is no difference. If g is symmetric, then there is absolutely no difference. If you use $g(-t)$ instead of g, then convolution becomes correlation. The only difference is a minor change in the Fourier theorems. Thus, we move to one of the most important theorems of Fourier Analysis.

Theorem 4.3.4 (Convolution Theorem) *If $f(t)$ and $g(t)$ are elements of $L^2(\mathbb{R})$, then the convolution $f * g(t)$ is also an element of $L^2(\mathbb{R})$. More importantly, the Fourier Transform of $f * g$ is given simply by $\hat{f}(s)\hat{g}(s)$, or pointwise multiplication of the corresponding Fourier Transforms. Mathematically, we have*

$$\mathcal{F}(f * g) = \hat{f}(s)\hat{g}(s). \tag{4.15}$$

This theorem states that the Fourier Transform diagonalizes all convolution operators. If convolution operators were rare, this wouldn't mean much.

Convolution operators are extremely common, however. Mathematically, any linear, time-independent operator is a convolution operator. Thus, if you have a machine, or process, which produces the same result from an input, regardless of the time the input is given, and if that machine is linear (most are), then you have a convolution process.

Proof: Most Fourier theorems are very methodical, so we begin with the definition of the Fourier Transform and 4.7.7

$$\mathcal{F}(f * g) = \frac{1}{\sqrt{2\pi}} \int f * g(t) e^{ist} dt = \frac{1}{\sqrt{2\pi}} \int \frac{1}{\sqrt{2\pi}} \int f(\tau) g(\tau - t) d\tau e^{ist} dt.$$

Moving the exponential inside of both integrations, we get

$$\mathcal{F}(f * g) = \frac{1}{2\pi} \int \int f(\tau) g(\tau - t) e^{ist} d\tau dt$$

$$= \frac{1}{2\pi} \int \int f(\tau) g(\tau - t) e^{-is(\tau - t)} e^{is\tau} d\tau dt.$$

Now, we can rearrange the order of integration which yields

$$\mathcal{F}(f * g) = \frac{1}{2\pi} \int f(\tau) \int g(\tau - t) e^{-is(\tau - t)} dt e^{is\tau} d\tau.$$

Concentrating on the inner integral, we see that it is independent of τ and we let $x = -(\tau - t)$ yielding

$$\mathcal{F}(f * g) = \frac{1}{\sqrt{2\pi}} \int f(\tau) \frac{1}{\sqrt{2\pi}} \int g(s) e^{is(s)} dx e^{is\tau} d\tau$$

$$= \frac{1}{\sqrt{2\pi}} \int f(\tau) e^{is\tau} d\tau \hat{g}(s)$$

$$= \hat{f}(s) \hat{g}(s).$$

Getting back to the difference between correlation and convolution, the reason for the $t - \tau$ in convolution rather than $\tau - t$ in correlation is simply that one gives a slightly cleaner theorem. The Correlation theorem is only slightly different, and the proof is nearly identical.

Theorem 4.3.5 (Correlation Theorem) *If $f(t)$ and $g(t)$ are elements of $L^2(\mathbb{R})$, then the correlation $f \star g(t)$ is also an element of $L^2(\mathbb{R})$. More importantly, the Fourier Transform of $f \star g$ is given simply by $\hat{f}(s)\overline{\hat{g}(s)}$, or pointwise multiplication of the corresponding Fourier Transforms $\hat{f}(s)$ and the conjugate $\overline{\hat{g}(s)}$. Mathematically, we have*

$$\mathcal{F}(f * g) = \hat{f}(s)\overline{\hat{g}(s)}. \tag{4.16}$$

A basic example:

Many physical systems can be modeled by a simple linear differential equation of the type

$$ay''(t) + by'(t) + cy(t) = r(t), \tag{4.17}$$

where $y(t)$ is the behavior of the system, a, b, and c are fixed constants, and $r(t)$ is the external forcing function, which can oftentimes be thought of as the input to the system. Now utilizing the Fourier Transform and its description of differentiation, we can change this to

$$-as^2\hat{y}(s) + b(is)\hat{y}(s) + c\hat{y}(s) = \hat{r}(s).$$

Solving algebraically, we now have $(-as^2 + ibs + c)\hat{y}(s) = \hat{r}(s)$, or

$$\hat{y}(s) = \left(\frac{1}{-as^2 + ibs + c}\right)\hat{r}(s) = \hat{T}(s)\hat{r}(s). \tag{4.18}$$

Notice a couple of things here. First, we have reduced solution of a linear differential equation to a simple algebraic formula in terms of the Fourier Transform. Second, that algebraic formula becomes a multiplication by $\hat{T}(s)$ in 4.18, and therefore, we have reduced equation 4.17 to a convolution equation

$$y(t) = \int r(\tau)T(\tau - t)d\tau.$$

The function $T(t)$ is oftentimes referred to as the transfer function of the operator, or operation.

We want to emphasize a number of things at this time. First, this example nicely shows that differential equations oftentimes become convolution equations, so differentiation and convolution are very closely linked. Secondly, the above analysis can be carried out for higher-order differential equations. Thus, we can describe any constant coefficient linear differential equation as a Fourier multiplier, or a convolution equation.

Your stereo:

We mentioned in the introduction that your stereo can be represented simply via the Fourier Transform. The above analysis is exactly the mathematics that does that. Specifically, the electrical circuit in a traditional stereo is an RLC circuit, with R standing for resistor, L standing for inductor, and C standing for capacitor. The characteristics of the RLC components will determine the coefficients a, b, and c. Thus, the nature of the output of the stereo is chosen appropriately to manipulate the transfer function $T(t)$, or its Fourier pair $\hat{T}(s)$.

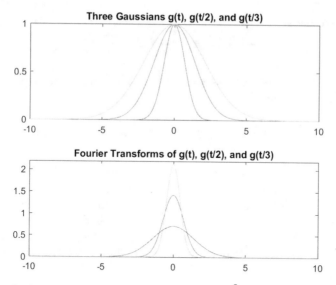

Figure 4.2:   In the top figure, we plot a Gaussian $g(t) = e^{-t^2}$ in blue, along with its dilation $g(t/2)$ in the red, and its dilation $g(t/3)$ in yellow. All of the functions are identical with the exception of dilation. At the bottom, we plot the Fourier Transforms of $g(t), g(t/2)$, and $g(t/3)$ once again in blue, red, and yellow. Note that the narrowest function in time, g(t), has the broadest Fourier Transform. Similarly, the broadest function in time, $g(t/3)$ has the narrowest Fourier Transform.

4.3.4   Dilation and Uncertainty

We begin with the scaling, or dilation, of a function in $L^2(\mathbb{R})$. We define this below.

Definition 4.3.4 (Dilation) *Let $f(t) \in L^2(\mathbb{R})$. We define refer to scaling, or dilation, of a function by a factor of a as $D_a f(t) = f(at)$.*

Thus, if $f(t)$ was a function which was non-zero on $[-1, 1]$, and $a = 2$, then $D_2 f(t) = f(2t)$ would be non-zero from $[-.5, .5]$, or it would be shrunk by a factor of 2. Similarly, if $a = 1/2$, then the function $f(t/2)$ would be stretched by a factor of 2. The question which we want to answer is "Given the Fourier Transform of $f(t)$, which we call $\hat{f}(s)$, what can we say about the Fourier Transforms of $f(2t)$, or $f(t/2)$, or more generally $f(at)$?" This, like many Fourier questions, is answered by simple Freshman Calculus.

Theorem 4.3.6 *Let $f(t) \in L^2(\mathbb{R})$ and let $\hat{f}(s)$ be its Fourier Transform. Then if the Fourier operator is denoted by $\mathcal{F}$, we have*

$$\mathcal{F}(D_a(f(t))) = \mathcal{F}(f(at)) = \frac{1}{a}\hat{f}\left(\frac{s}{a}\right).$$

Proof: We have

$$\mathcal{F}(f(at)) = \frac{1}{\sqrt{2\pi}} \int f(at) e^{ist} dt.$$

We simply want to change variable on the right hand side, letting $x = at$ so $dx = a\,dt$, and we have

$$\mathcal{F}(f(at)) = \frac{1}{\sqrt{2\pi}} \int f(s) e^{i(s/a)x} dx/a = \frac{1}{a} \hat{f}(s/a).$$

Mathematically this is easy, but we should stop to understand this intellectually. If you consider Theorem 4.7.9, think of the sine and cosine terms involved in the first function. Those same sine and cosine terms will represent the second function, but they will be at double the frequency, since they will only last half as long. Similarly, the terms from the first function will be elongated by a factor of two in the third function, or at $1/2$ the frequencies.

Uncertainty:

We now move on to the idea of uncertainty, or the uncertainty principle. In quantum mechanics, this involves the inability to know the precise location and velocity of an electron at a certain point in time. In Fourier Analysis, this means that a function cannot be extremely short in time, and extremely short in frequency. The Dilation theorem 4.7.9 states that if a function is contracted in time, it must be expanded in frequency. Let us try to define this via the statistical idea of variance, or the second moment of a function. But first, we must think about dilation in a uniform way, i.e., in a way that doesn't change the norm of the function.

Recall from Theorem 4.7.9, we have

$$\mathcal{F}(f(at)) = \frac{1}{a} \hat{f}\left(\frac{s}{a}\right).$$

The problem with this is that the norm $\|f(at)\|_2$ will change with a. What we want is to dilate f without changing its norm.

Definition 4.3.5 (Uniformly dilated pairs) *Let $f(t)$ be any element of $L^2(\mathbb{R})$. Then we define the uniformly dilated pairs for f to be*

$$\sqrt{a} f(at) \quad and \quad \frac{1}{\sqrt{a}} \hat{f}\left(\frac{s}{a}\right).$$

A simple calculation shows that

$$\|\sqrt{a} f(at)\|_2^2 = \int a|f(at)|dt = \int |f(s)|dx = \|f(t)\|_2^2,$$

where we have made the simple substitution $x = at$. By the isometry, we know that the norm of $\|\hat{f}(s)\| = \|f(t)\|$, so this type of normalized dilation preserves the energy in the system.

We now want to consider the product of the variance of a function, and the variance of its Fourier Transform. Namely

$$\|tf(t)\|_2 \|s\hat{f}(s)\|_2.$$

We want to go further and show that this is independent of uniform dilation, or independent of a in the uniform dilation.

Theorem 4.3.7 *The quantity*

$$\int t^2 |f(t)|^2 dt \int s^2 |\hat{f}(s)|^2 ds \tag{4.19}$$

is independent of uniform dilation. In other words,

$$\int t^2 a |f(at)|^2 dt \int s^2 \frac{1}{a} \left| \hat{f}\left(\frac{s}{a}\right) \right| ds$$

is a constant.

Proof: This is once again very straight forward. Beginning with

$$\int t^2 a |f(at)|^2 dt \int s^2 \frac{1}{a} \left| \hat{f}\left(\frac{s}{a}\right) \right| ds,$$

we substitute $x = at$ and $ua = s$, and we get

$$\int \frac{x^2}{a^2} a |f(s)|^2 \frac{1}{a} dx \int (ua)^2 \frac{1}{a} |\hat{f}(u)| a ds.$$

$$= \int x^2 |f(s)|^2 dx \int u^2 |\hat{f}(u)| ds.$$

Thus, uniform dilation does not change the quantity in 4.19.

While this lets us know that dilation does not change the product 4.19, we have not shown what this product is. How small can it be for any arbitrary choice of f? Is there an optimal f? The answer is given in the classical theorem.

Theorem 4.3.8 (Uncertainty Principle) *Let $f(t)$ be a real valued function and $\hat{f}(s)$ be in $L^2(\mathbb{R})$, and let $tf(t)$ and $s\hat{f}(s)$ also be in $L^2(\mathbb{R})$. Assume for simplicity that $tf^2(t) \to 0$. Then we have*

$$\int t^2 |f(t)|^2 dt \int s^2 |\hat{f}(s)|^2 ds \geq \frac{1}{4} \|f(t)\|^4. \tag{4.20}$$

If $\|f(t)\| = 1$, then we have the simpler form,

$$\int t^2 |f(t)|^2 dt \int s^2 |\hat{f}(s)|^2 ds \geq \frac{1}{4}. \tag{4.21}$$

In addition, equality is only obtained by scaled dilates of the Gaussian, or normal distribution function e^{-t^2}.

Proof: Once again we move methodically. First, remember that $f'(t)$ exists and has the Fourier Transform $s\hat{f}(s)$. Then we can use the Fourier isometry to get

$$\int t^2 |f(t)|^2 dt \int s^2 |\hat{f}(s)|^2 ds = \int t^2 |f(t)|^2 dt \int |f'(t)|^2 dt \tag{4.22}$$

Now we utilize the Cauchy Schwartz inequality stating that $\|a\|^2 \|b\|^2 \geq |\langle a, b \rangle|^2$ to get

$$\int t^2 |f(t)|^2 dt \int s^2 |\hat{f}(s)|^2 ds \geq \left| \int t f(t) f'(t) dt \right|^2 \tag{4.23}$$

$$= \left| \frac{1}{2} \int t \frac{d}{dt} [f^2(t)] dt \right|^2.$$

Now remembering that $\int fg' = fg - \int f'g$, we have,

$$\int t^2 |f(t)|^2 dt \int s^2 |\hat{f}(s)|^2 ds \geq \left| \frac{1}{2} \int t \frac{d}{dt} [f^2(t)] dt \right|^2$$

$$= \frac{1}{4} \left| t f^2(t) \Big|_{-\infty}^{\infty} - \int f^2(t) dt \right|^2.$$

Now since $tf(t) \in L^2(\mathbb{R})$, it follows that $tf^2(t) \to 0$, so we have

$$\int t^2 |f(t)|^2 dt \int s^2 |\hat{f}(s)|^2 ds \geq \frac{1}{4} \|f(t)\|^4. \tag{4.24}$$

Notice that we have only one inequality, and that is due to the Cauchy Schwartz inequality in 4.23. Equality can only be obtained in the Cauchy Schwartz inequality $|\langle f, g \rangle| \leq \|f\| \|g\|$ if $f = kg$ or in this case if $f'(t) = ctf(t)$. We recognize this as a simple separable differential equation leave the result as an exercise. Thus, theorem is proven. $\qquad \square$

The reader should not become alarmed upon seeing this theorem with a slightly different constant at the right. If the definition of the Fourier Transform is changed slightly, then the constant changes. The absolute constant rarely becomes an issue. The general idea is very important. Stated in English, *"If a function is very well localized in time, it cannot be very well localized in frequency, and vice versa."*

4.3.5   The Fourier Transform of a Gaussian

We will now move toward that by examining a very special function for all of mathematics. The Gaussian function is central to statistics, differential equations, and Fourier Analysis. For any $a > 0$, we define a basic Gaussian function to be

$$g_a(t) = e^{-at^2}.$$

We now want to take its Fourier Transform. We will follow the work of [10] as was presented in [5]. By definition, we have that

$$\hat{g}_a(s) = \frac{1}{\sqrt{2\pi}} \int_{-\infty}^{\infty} e^{-at^2} e^{ist} dt.$$

We also know that we can differentiate this with respect to s and get

$$\frac{d}{ds}(\hat{g}_a(s)) = \hat{g}_a'(s) = \frac{i}{\sqrt{2\pi}} \int_{-\infty}^{\infty} t e^{-at^2} e^{ist} dt. \tag{4.25}$$

Now we know that the derivative of $\exp(-at^2)$ is $-2at \exp(-at^2)$, so we can rewrite 4.25 as

$$\hat{g}_a'(s) = \frac{-i}{2a\sqrt{2\pi}} \int_{-\infty}^{\infty} \frac{d}{dt}(e^{-at^2}) e^{ist} dt. \tag{4.26}$$

If we integrate 6.12 by parts, we get

$$\hat{g}_a'(s) = \frac{-i}{2a\sqrt{2\pi}} \left[(e^{-at^2}) e^{ist} \right]_{-\infty}^{\infty} - \frac{-i}{2a\sqrt{2\pi}} \int_{-\infty}^{\infty} (e^{-at^2})(is) e^{ist} dt$$

$$= 0 + \frac{-s}{2a\sqrt{2\pi}} \int_{-\infty}^{\infty} (e^{-at^2}) e^{ist} dt = \frac{-s}{2a} \hat{g}(s). \tag{4.27}$$

Thus, we have a simple first-order linear differential equation

$$\hat{g}'(s) = \frac{-s}{2a} \hat{g}(s),$$

for which there is only one linearly independent solution,

$$\hat{g}(s) = k e^{-s^2/(4a)}.$$

This is a remarkable result. The Fourier Transform of a Gaussian is also a Gaussian. We obviously know from the scaling theorem that if the Gaussian is narrow, the Fourier Transform will be wide, and vice versa. This can be seen from this result. Namely the Fourier tranform of $\exp(-at^2)$ is $k \exp(-s^2/(4a))$, so the reciprocal scaling factor is present.

To determine the constant k, we need only remember from the definition of the Fourier Transform $\hat{g}(0) = \frac{1}{\sqrt{2\pi}} \int g_a(t) e^{i0t} dt = \frac{1}{\sqrt{2\pi}} \int e^{-at^2} dt = k$. Using the standard formula $\int e^{-u^2} du = \sqrt{\pi}$, the change of variable $w = \sqrt{a} t$ gives us

$$k = \frac{1}{\sqrt{2\pi}} \int e^{-at^2} dt = \frac{1}{\sqrt{2\pi}} \frac{1}{\sqrt{a}} \int e^{-w^2} dw = \frac{\sqrt{\pi}}{\sqrt{2\pi a}} = \frac{1}{\sqrt{2a}}.$$

Thus, we have

$$\hat{g}_a(s) = \frac{1}{\sqrt{2a}} e^{-s^2/(4a)},$$

or we can use the notation

$$e^{-at^2} \Leftrightarrow \sqrt{\frac{1}{2a}} e^{-s^2/(4a)}.$$

Note that by symmetry, or dilation, we also have

$$\sqrt{\frac{1}{2a}} e^{-t^2/(4a)} \Leftrightarrow e^{-as^2}, \tag{4.28}$$

which we leave as exercise 1 below.

4.3.6 Problems and Exercises:

1. Show using either dilation or symmetry that 4.28 is also valid.
   General Relationships For the following exercises, let us assume that we have an arbitrary function $f(t) \in L^2(\mathbb{R})$ with a corresponding Fourier Transform $\hat{f}(s)$.

2. Find the Fourier Transform of $f(at - b)$, in terms of $\hat{f}(s)$.

3. Find the Fourier Transform of $f(a(t - b))$, in terms of $\hat{f}(s)$. Is this different than the problem above?

4. Find the Fourier Transform of $\frac{d}{dt}(f(at - b))$, in terms of $\hat{f}(s)$.

5. Find the Fourier Transform of $\frac{d}{dt}(f(a(t-b)))$, in terms of $\hat{f}(s)$.

6. Find the Fourier Transform of $\frac{d}{dt}(f(t)g(t))$, in terms of $\hat{f}(s)$ and $\hat{g}(s)$.

7. Find the Fourier Transform of $\frac{d}{dt}(f(t)*g(t))$, in terms of $\hat{f}(s)$ and $\hat{g}(s)$.

4.4   The Inverse Transform

4.4.1   Proving the Fourier Inversion Formula

We defined the Inverse Fourier Transform, but did not prove that it was valid. We have shown that the Fourier Transform is an isometry, which would generally mean that it is invertible. We have not, however, shown that the simple formula which we put forth is the inversion formula. We restate this at this time.

Theorem 4.4.1 (The Inverse Fourier Transform and Representation) *If $f(t) \in L^2(\mathbb{R})$, then its Fourier tranform $\hat{f}(s) \in L^2(\mathbb{R})$, and we can represent $f(t)$ by its Fourier Transform, or*

$$f(t) = \frac{1}{\sqrt{2\pi}} \int_{-\infty}^{\infty} \hat{f}(s)e^{-ist}ds. \qquad (4.29)$$

Proof: We have already proven the isometry, so all that is left is the proof of the actual inversion formula. Equals in the formula will be in the $L^2(\mathbb{R})$ sense, not in a pointwise sense. We will approximately follow along the lines presented in [6].

As we stated, when we defined the Fourier Transform, it is hard to see pointwise existence unless $f \in L^1(\mathbb{R})$. We will lean on L^1 to get the L^2 proof.

We will consider the convolution of $f(t)$ and $g_a(t) = \sqrt{\frac{\pi}{a}}e^{-t^2/(4a)}$. Recall from above that the Fourier Transform of $g_a(t) = \sqrt{\frac{\pi}{a}}e^{-t^2/(4a)}$ is e^{-as^2}. Thus, we consider the function

$$f(t) * g_a(t) = \frac{1}{\sqrt{2\pi}} \int f(s)g_a(t-x)dx.$$

The Convolution theorem tells us that

$$\mathcal{F}(f(t) * g_a(t)) = \hat{f}(s)\hat{g}_a(s) = \hat{f}(s)e^{-as^2}$$
$$= \hat{f}(s) \int \sqrt{\frac{\pi}{a}}e^{-t^2/(4a)}e^{ist}dt, \qquad (4.30)$$

where the last line is due to knowing the Fourier Transform of the Gaussian. Note that $\hat{f}(s)e^{-as^2} \in L^1(\mathbb{R})$, even if $\hat{f}(s) \notin L^1(\mathbb{R})$ (exercise). Thus, the

Inverse Fourier Transform of 4.30 will be well defined. We now calculate the
Inverse Fourier Transform

$$
\begin{aligned}
\mathcal{F}^{-1}(\hat{f}(s)e^{-as^2}) &= \frac{1}{\sqrt{2\pi}} \int \hat{f}(s)e^{-as^2}e^{-ist}\,ds \\
&= \frac{1}{\sqrt{2\pi}} \int \frac{1}{\sqrt{2\pi}} \int f(s)e^{ixs}\,dx\, e^{-as^2}e^{-ist}\,ds \\
&= \frac{1}{\sqrt{2\pi}} \int f(s)\frac{1}{\sqrt{2\pi}} \int e^{-as^2}e^{-i(t-x)s}\,ds dx \\
&= \frac{1}{\sqrt{2\pi}} \int f(s)\left(\sqrt{\frac{\pi}{a}}e^{-(t-x)^2/(4a)}\right)dx \\
&= \frac{1}{\sqrt{2\pi}} \int f(s)g_a(t-x)\,dx. && (4.31)
\end{aligned}
$$

Note that above, we only used the fact that we can directly calculate the
Fourier and Inverse Fourier Transform of the Gaussian, which is well defined.
   We want to consider this for $a \to 0$. Note that $\lim_{a\to 0} e^{-as^2} = 1$ for all
fixed s. Moreover, we have that $\lim_{a\to 0} e^{-(t-x)^2/(4a)} = 0$ for all fixed $t-x \neq 0$.
We also know that

$$
\frac{1}{\sqrt{2\pi}} \int \sqrt{\frac{\pi}{a}}e^{-(t-x)^2/(4a)}\,dx = 1
$$

for all t and a (exercise). The values of $\sqrt{\frac{\pi}{a}}e^{-(t-x)^2/(4a)}$ become extremely
centered about $t = x$, and the average of those values is always 1. Thus, the
convolution 4.31 will only consider averages of values of $f(s)$ very close to t.
In the limit, this implies that

$$
\lim_{a\to 0} \mathcal{F}^{-1}(\hat{f}(s)e^{-as^2}) = \lim_{a\to 0} \frac{1}{\sqrt{2\pi}} \int f(s)g_a(t-x)\,dx = f(t), \qquad (4.32)
$$

whenever $f(t)$ is continuous. We also have that $\hat{f}(s)e^{-as^2} \to \hat{f}(s)$. This
implies that

$$
\lim_{a\to 0} \mathcal{F}^{-1}(\hat{f}(s)e^{-as^2}) = \lim_{a\to 0} \frac{1}{\sqrt{2\pi}} \int \hat{f}(s)e^{-as^2}e^{-ist}\,dt = \frac{1}{\sqrt{2\pi}} \int \hat{f}(s)e^{-ist}\,dt.
$$

Thus, we have the inversion formula

$$
\mathcal{F}^{-1}(\hat{f}(s)) = \frac{1}{\sqrt{2\pi}} \int \hat{f}(s)e^{-ist}\,ds = f(t) \qquad (4.33)
$$

when $f(s)$ is continuous at t.
   The proof from here proceeds by showing that the above-stated conditions
make the equality hold in an $L^2(\mathbb{R})$ sense, as $a \to 0$, for f which are piecewise

continuous. The result is then extended to all of $L^2(\mathbb{R})$ by utilizing the fact that piecewise continuous functions are dense in $L^2(\mathbb{R})$. These finishing details are left as exercises.

□

4.4.2   Problems and Exercises:

1. Show that the Fourier Transform of a Gaussian, which is again a Gaussian as we have shown, has the same norm.

2. Show using either dilation or symmetry that 4.28 is also valid.

3. Calculate the Fourier Transform of $e^{-|t|}$.

4. Prove that if $\hat{f}(s) \in L^2(\mathbb{R})$ then $\hat{f}(s)e^{-as^2} \in L^1(\mathbb{R})$ for any $a > 0$.

5. Show that

$$\frac{1}{\sqrt{2\pi}} \int \sqrt{\frac{\pi}{a}} e^{-u^2/(4a)} du = 1,$$

by calculating a Fourier Transform at 0.

6. [**Analysis Background Required**] Suppose that $f(t) \in L^2(\mathbb{R})$, and that $f(t)$ is continuous about t_0. Give a rigorous proof of equation 4.32

7. Suppose that $f(t) \in L^2(\mathbb{R})$, and that $f(t)$ has a jump discontinuity at t_0. Suppose further that there is a neighborhood $N = [t_0 - \delta, t_0 + \delta]$ such that $f(t)$ is continuous in N with the exception of t_0. Let $f(t_0^+) = lim_{\epsilon \to 0} f(t + \epsilon)$ and $f(t_0^-) = lim_{\epsilon \to 0} f(t - \epsilon)$. Prove that

$$\mathcal{F}^{-1}(\hat{f}(s)) = \frac{1}{2}(f(t_0^+) + f(t_0^-)).$$

Do this by using the approach in the Inverse Fourier Transform. Specifically, examine the convolution equation 4.31.

8. [**Analysis Background Required**] Give a rigorous finish to the proof of Theorem 4.4.1 by utilizing the standard convergence theorems in appropriate ways. Problem 7 also guides the way.

4.5   Partition of Unity

The Fourier Transform gives us a very powerful tool to analyze functions. In addition, the Fourier Transform gives us a very powerful tool which allows us to analyze, design, and understand other orthonormal systems.

The question now is "What does the Fourier Transform have to say about another arbitrary orthonormal basis?" The answer is

Theorem 4.5.1 (Partition of Unity) *Suppose that* $\{\phi_n(t)\}$ *is an ortho-normal basis for* $L^2[a,b]$. *Then*

$$\frac{2\pi}{b-a} \sum_n |\hat{\phi}_n(s)|^2 = 1.$$

Proof: The proof is very straight forward, as are many Fourier Analysis proofs. This emphasizes the value of Fourier Analysis as a tool.

We will simply analyze the sum in two different ways. Suppose that $\{\phi_n\}$ is an arbitrary orthonormal sequence on $L^2[a,b]$. Then by the Fourier Isometry,

$$\|e^{ist}\|^2 = \sum_n |\langle \phi_n, e^{ist} \rangle|^2$$

$$= \sum_n \left| \int \phi_n(t) e^{ist} dt \right|^2$$

$$= 2\pi \sum_n \left| \frac{1}{\sqrt{2\pi}} \int \phi_n(t) e^{ist} dt \right|^2$$

$$= 2\pi \sum_n |\hat{\phi}_n(s)|^2. \qquad (4.34)$$

Recalling that

$$\|e^{ist}\|^2 = \int_a^b |e^{ist}|^2 dt = \int_a^b 1 \, dt = (b-a),$$

we have that

$$\|e^{ist}\|^2 = 2\pi \sum_n |\hat{\phi}_n(s)|^2 = (b-a).$$

Combining these gives us

$$\|e^{ist}\|^2 = (b-a) = 2\pi \sum_n |\hat{\phi}_n(s)|^2,$$

which proves the result.   □

It is very informative to think through this theorem. This means that any orthonormal system is just a partitioning of the frequency spectrum. Thus, if you choose orthogonal polynomials or any other basis, you are choosing another way to partition the spectrum. There are many reasons to choose an orthonormal system, but the above Theorem gives all of them a common basis in Fourier Analysis. In addition, the above analysis allows us to design orthonormal systems at times.

4.6 Revisiting Gibbs' ringing

We will now revisit Gibbs' ringing from the viewpoint of the continuous Fourier Transform. While Gibbs' ringing is generally thought of as a phenomena which happens when the partial sums of a Fourier Series converges, the continuous Fourier Transform allows us to explain it more easily. We will follow the approach in [5], where an excellent historical discussion of the phenomenon may also be found.

We will begin by assuming that $f(t)$ is a function in $L^2(\mathbb{R})$. We will also assume that $f(t)$ has a jump discontinuity, which implies that its Fourier Transform does not decay quickly. We want to see what happens when we restrict its Fourier Transform to a finite interval, or consider $\hat{h}(s) = \hat{f}(s)\chi_\Omega(s)$. The convolution theorem therefore implies that we are looking at the function

$$h(t) = f(t) * \mathcal{F}^{-1}(\chi_\Omega(s)).$$

A quick calculation (exercise 2) will show that

$$\mathcal{F}^{-1}(\chi_\Omega(s)) = \sqrt{\frac{2}{\pi}}\frac{\sin(\Omega t)}{\Omega t}, \qquad (4.35)$$

so we have the convolution

$$\begin{aligned} f * \frac{\sin(\Omega t)}{\Omega t} &= \frac{1}{\sqrt{2\pi}}\int f(s)\sqrt{\frac{2}{\pi}}\frac{\sin(\Omega(t-x))}{\Omega(t-x)}dx \\ &= \int f(s)\frac{\sin(\Omega(t-x))}{\pi\Omega(t-x)}dx. \end{aligned} \qquad (4.36)$$

For clarity of understanding, let us assume that $f(t) = 1$ for $t > 0$ and $f(t) = 0$ for $t < 0$. Now $f(t) \notin L^2(\mathbb{R})$, but the convolution 4.36 is well defined and will converge. We think of $f(t)$ as a limiting case, which allows us to study the phenomena more easily. Now since $\chi_\Omega(0) = 1$, it follows that

$$\int \frac{\sin(\Omega t)}{\pi\Omega t}dt = 1.$$

By the nature of f, we have that 4.36 is equal to

$$f * \frac{\sin(\Omega t)}{\Omega t} = \int_{x=0}^{\infty}\frac{\sin(\Omega(t-x))}{\Omega(t-x)}dx = \int_{x=0}^{\infty}\frac{\sin(\Omega(x-t))}{\Omega(x-t)}dx, \qquad (4.37)$$

where the last step on the right uses the fact that $\sin(t)/t$ is even. We can rewrite this simply as

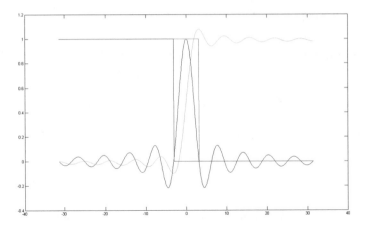

Figure 4.3: We illustrate the function $\sin(t)/t$ above. Note that a maximum value of the integral 4.38 will occur where the integral is denoted by the red curve, and a minimum denoted by the green curve. The fourth curve is the convolution. Note that it tends to 1, for $t \ll 0$ and it approaches zero for $t \gg 0$. This is consistent with what we know.

$$\int_{x=t}^{\infty} \frac{\sin(\Omega x)}{\pi \Omega x} \, dx. \tag{4.38}$$

We have illustrated this convolution, and plotted the convolution in Figure 4.3.

To understand Figure 4.3, we recall that the integrand in 4.38 is even, so we have

$$\int_{-\infty}^{0} \frac{\sin(\Omega x)}{\pi \Omega x} \, dx = \int_{0}^{\infty} \frac{\sin(\Omega x)}{\pi \Omega x} \, dx = \frac{1}{2}.$$

We now want to understand the values at the two peaks, which is what we consider to be "Gibb's Ringing". We consider the closeup of Figure 4.3 in Figure 4.4.

We see that we can write the value of

$$\int_{-\infty}^{t} \frac{\sin(\Omega x)}{\pi \Omega x} \, dx = \int_{-\infty}^{0} \frac{\sin(\Omega x)}{\pi \Omega x} \, dx \pm \int_{0}^{t} \frac{\sin(\Omega x)}{\pi \Omega x} \, dx$$

$$= \frac{1}{2} \pm \int_{0}^{t} \frac{\sin(\Omega x)}{\pi \Omega x} \, dx, \tag{4.39}$$

where the $\pm$ is exactly the sign of t. Thus, the maximum and minimums will be

$$\frac{1}{2} \pm \int_{0}^{M} \frac{\sin(\Omega x)}{\pi \Omega x} \, dx,$$

where M is the value that maximizes the integral. It is fairly easy to see that M occurs at the first zero of $\sin(\Omega x)$, or when $\Omega x = \pi$. Thus, the maxima and minima occur at

$$M = \pm \frac{\pi}{\Omega}.$$

Thus, the maximum and minimum values are determined by

$$\int_0^{\pi/\Omega} \frac{\sin(\Omega x)}{\pi \Omega x} dx.$$

Using the change of variables $t = \Omega x$, we get that this is

$$\frac{1}{2} \pm \int_0^{\pi} \frac{\sin t}{\pi t} dt, \tag{4.40}$$

which is independent of Ω! This means that now matter how many frequency terms are included, there will still be some overshoot. Thus, the magnitude of the overshoot is independent of Ω.

This seems to contradict the fact that the Fourier Transform will converge. The reality is that is doesn't converge uniformly. In addition, it will not always converge pointwise, but in an average sense of $L^2(\mathbb{R})$.

The question remains, how big is 4.40. It is hard to calculate in closed form, but one can easily come up with a series by using the Taylor approximation. Let us just say that the maximum is greater than 1, and the minimum is less than 1, which supports our numerics. The answer is left as an exercise. As you can see, however, from the illustration, the maximum is somewhere around 1.09. Use this to guide your calculations in exercise 3.

4.6.1 Problems and Exercises:

1. **Partion of Unity** We know that $f_k(t) = \frac{1}{2\pi} e^{ikt} \chi_\pi(t)$ is an orthonormal basis for $L^2[-\pi, \pi]$. Verify the Partition of Unity by (a) Calculating the Fourier Transforms $\hat{f}_k(s)$, by using translation relationship, or directly. (b) Plot the sum of squares

$$\sum_{k=-n}^{n} |\hat{f}_k(s)|^2$$

in MATLAB, where n is perhaps 5. This will not prove the theorem but should give you an idea of the way the sum converges.

2. Calculate $\mathcal{F}^{-1}(\chi_\Omega(s))$.

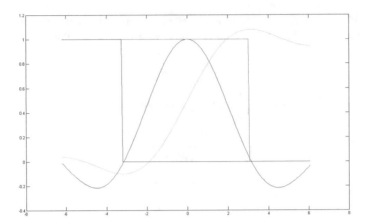

Figure 4.4: We illustrate the function $\sin(t)/t$ above. Note that a maximum value of the integral 4.38 will occur the integral is denoted by the red curve, and a minimum denoted by the green curve. The fourth curve is the convolution. Note that the maximum values and minimum values occur at $\pm\pi$.

3. Calculate the Gibb's constant from 4.40. Utilize the Taylor series for $\sin t$ and integrate term by term. The final answer can be calculated numerically. Try to make sure you have enough terms to be accurate to 8 terms. Generally the Gibb's constant is referred to as the amount over 1, or below zero, attained by 4.40. This is the amount over or under the desired value.

4. Return to the Fourier Series calculations you did for $\chi_{\pi/2}(t)$ on $[-\pi, \pi]$. We have claimed above that the maximum point for the Gibbs ringing is $M = \pi/\Omega$. This implies that the point moves to the discontinuity as π/Ω. Check out this claim with the Fourier Series. You many need many points. If you double the number of Fourier coefficients, what happens to the maximum points? Are they moving toward the discontinuity? Record these points, calculate the distance to the discontinuity, $\pi/2$, and graph them. Are they consistent with our analysis? If you record the distances to the discontinuity, it may be helpful to plot 1 over that distance. This graph should look like a line. Why?

4.7   Suggested Test Review:

Generally, my one test looks similar to this. I give them an enlarged version of the test, so they can study. Obviously, being able to prove all of the "Hard One" is a bit much for undergraduates. I usually require 5 to be graded. A 100 is 60 points a very mature student can get 75 points. That's rare just because of time constraints.

Undoubtably, each individual Professor will choose a different assortment of problems. My purpose is so they have the basic theorems memorized, if not all of the "hard" proofs.

The Hard Ones: 15 pts

Theorem 4.7.1 (Fourier Isometry) *Let $f(t) \in L^2(\mathbf{R})$, and let $\hat{f}(s)$ be its continuous Fourier Transform as defined in Definition 4.1.4. Then we have*

$$\int_{-\infty}^{\infty} |f(t)|^2 dt = \int_{-\infty}^{\infty} |\hat{f}(t)|^2 dt. \tag{4.41}$$

In addition, the Fourier Transform is an isometry from $L^2(\mathbf{R})$ to $L^2(\mathbf{R})$.

Theorem 4.7.2 (Uncertainty Principle) *Let $f(t)$ and $\hat{f}(s)$ be in $L^2(R)$, and let $tf(t)$ and $s\hat{f}(s)$ also be in $L^2(R)$. Assume for simplicity that $tf^2(t) \to 0$. Then we have*

$$\int t^2 |f(t)|^2 dt \int s^2 |\hat{f}(s)|^2 ds \geq \frac{1}{4} \|f(t)\|^4. \tag{4.42}$$

If $\|f(t)\| = 1$, then we have the simpler form,

$$\int t^2 |f(t)|^2 dt \int s^2 |\hat{f}(s)|^2 ds \geq \frac{1}{4}. \tag{4.43}$$

In addition, equality is only obtained by scaled dilates of the Gaussian, or normal distribution function e^{-t^2}.

Theorem 4.7.3 (Partition of Unity) *Suppose that $\{\phi_n(t)\}$ is an orthonormal basis for $L^2[a, b]$. Then*

$$\frac{1}{b-a} \sum_n |\hat{\phi}_n(s)|^2 = 1.$$

Theorem 4.7.4 (The Shannon Sampling Theorem)
Let $f(t) \in L^2(-\infty, \infty)$ and suppose that $\hat{f}(s) = 0$ for $|s| > \pi$. Then

$$f(t) = \sum_k f(k) \frac{\sin(\pi(t-k))}{\pi(t-k)}.$$

Theorem 4.7.5 (Fourier Transform of a Gaussian) *The Fourier Transform of a Gaussian is once again a Gaussian.*

The Easy Ones, 10 pts:

Theorem 4.7.6 (Differentiation) *Let* $f(t) \in L^2(\mathbf{R})$ *and let* $\hat{f}(s) \in L^2(\mathbf{R})$ *be its Fourier Transform. If* $(-is)\hat{f}(s) \in L^2(\mathbf{R})$, *then it follows that* $f'(t)$ *exists as an element of* $L^2(\mathbf{R})$, *and that its Fourier Transform is*

$$\mathcal{F}(f'(t)) = (-is)\hat{f}(s).$$

Theorem 4.7.7 (Convolution Theorem) *If* $f(t)$ *and* $g(t)$ *are elements of* $L^2(R)$, *then the convolution* $f * g(t)$ *is also an element of* $L^2(R)$. *More importantly, the Fourier Transform of* $f * g$ *is given simply by* $\hat{f}(s)\hat{g}(s)$, *or pointwise multiplication of the corresponding Fourier Transforms. Mathematically, we have*

$$\mathcal{F}(f * g) = \hat{f}(s)\hat{g}(s). \tag{4.44}$$

Theorem 4.7.8 (Translation) *If* $f(t) \in L^2(R)$ *and* $\hat{f}(s)$ *is its Fourier Transform, then*

$$\mathcal{F}(T_a(f(t)) = \mathcal{F}(f(t - a)) = \hat{f}(s)e^{isa}. \tag{4.45}$$

Theorem 4.7.9 (Dilation) *Let* $f(t) \in L^2(R)$ *and let* $\hat{f}(s)$ *be its Fourier Transform. Then if the Fourier operator is denoted by* $\mathcal{F}$, *we have*

$$\mathcal{F}(D_a(f(t)) = \mathcal{F}(f(at)) = \frac{1}{a}\hat{f}\left(\frac{s}{a}\right).$$

1. Show that if f is even, $\hat{f}(s)$ is real and symmetric, and if f is odd, $\hat{f}(s)$ is purely imaginary and antisymmetric.

2. Show that if $f(t) \geq 0$, then the maximum value of $\hat{f}(s)$ is attained by $\hat{f}(0)$.

3. Show how the differential equation $ay'' + by' + cy = f(t)$ can be written as a convolution problem.

Chapter 5
Sampling and Interpolation

We will consider one of the most fundamental applications of Fourier analysis. If you turn on your stereo, listen to the radio, talk on your cell phone, or use the internet today, you will be utilizing the fundamental applications of this chapter. Sampling and interpolation are fundamental to modern life.

We have already talked somewhat about sampling in Chapter 4. We mentioned that the coefficients of the Fourier Series of a function on $[-T, T]$ are really just samples of the continuous Fourier Transform, as shown in 4.5. The idea of taking a countable number of samples from the continuous Fourier Transform is the essence of sampling. Often times, however, we are not sampling the continuous Fourier Transform $\hat{f}(s)$, but rather the continuous time representation of a function $f(t)$. The fundamental question is

> **The Sampling Question:** Can I exactly represent a continuous function from a finite, or countably infinite number of samples of that function $\{f(t_k)\}_{-\infty}^{\infty}$? If not, can I approximately represent that function, with subsamples of the function?

The answer is yes, sometimes, and almost always if you know what you are doing. The practical use of the sampling theorems which will be presented in this chapter is almost always a matter of approximations. In reality, the most fundamental theorem of this chapter, The Shannon Sampling Theorem, never meets the requirement to use it. So what good is it? It works as a very good approximation to the truth. It gives the fundamental ideas for utilizing it, although they are almost never exactly met. The most important thing is to be close enough for the application. While the theorems suggest exact equalities, they are never exact in reality. Understanding this gray area is the art, and science, of sampling.

© Springer Science+Business Media, LLC 2017
T. Olson, *Applied Fourier Analysis*,
https://doi.org/10.1007/978-1-4939-7393-4_5

5.1 The Shannon Sampling Theorem

The Shannon Sampling Theorem could be attributed to a number of people before Shannon. Regardless of its academic origins, the theorem is most generally attributed to Shannon, who championed its applied use. We begin with a definition which is central to this theorem.

Definition 5.1.1 *A function $f(t) \in L^2(\mathbb{R})$ is said to be Ω-band limited if its Fourier Transform $\hat{f}(s) = 0$ for all $|s| > \Omega$.*

The Shannon Theorem is valid for these types of functions.

Theorem 5.1.1 (Shannon Sampling Theorem) *Let $f(t)$ be a function in $L^2(\mathbb{R})$, and let $\hat{f}(s)$ be its Fourier Transform. If $\hat{f}(s) = 0$ for $|s| > \pi$, then we can represent $f(t)$ exactly by the values of $f(t)$ at the integers $f(k)$, and the representation is given by*

$$f(t) = \sum_k f(k) \frac{\sin(\pi(t-k))}{\pi(t-k)} \tag{5.1}$$

where $k = -\infty \cdots -2, -1, 0, 1, 2, \ldots \infty$. Thus, you can exactly represent $f(t)$ by its values at the integers.

Before we prove the theorem, let us consider its uses and ramifications. It is hard to overstate exactly how intrinsic to modern science and engineering, and also to average modern life this theorem is. This theorem and its improvements are embedded in everyday life. You might think you are the only one speaking on the particular frequency that your cell phone is using. In reality, hundreds or perhaps thousands of people are speaking at the same time on that one channel. The human voice only has a certain amount of bandwidth, so a few samples taken rapidly allow one to exactly reconstruct not only the words, but the voice character of the conversation. When you speak into your phone, it is sampled, digitized, transmitted, received, and reconstructed all in real time. It is so accurate that we generally can recognize our children, friends, and parents even though the actual voice signal is never sent. If there are 100 people on that particular frequency, the samples of your voice are transmitted in every 100'th bit. The receiving cell phone knows which bits to grab and then reconstructs your voice.

If you have cable TV, the reason you have perhaps 200 channels is that the whole signal you see is not transmitted. A great deal of effort has been made to sample, digitize, and reconstruct the channels, as quickly and accurately as possible. You may have been in an institution with both digital and non-digital TV transmissions of the same event. If it was an athletic event, it probably was fairly confusing, because the non-digital signal will generally show up 3–5 seconds before the digitized signal. The digital high definition picture is better, but the low-resolution channel does not need coding and decoding. As a result, the low-resolution TV channel will show a touchdown, goal, or basket 3–5 seconds earlier because of encoding and decoding time. Its confusing in a sports bar, because different parts of the crowd erupt at different times depending on which TV they are watching. It leaves one conflicted between the higher resolution channel and the quicker one. These are real engineering questions.

We could go on with other applications. The reality is that it is very unlikely that anyone in our society does not encounter the use of sampling theorems during any one day. Let us prove the theorem.

Proof of Theorem 5.1.1: The proof of the theorem, as is the case with most proofs in Fourier analysis, is rather straight forward. It may seem somewhat backward, however, since we begin by expressing the Fourier Transform of $f(t)$, $\hat{f}(s)$, in its standard Fourier Series. Thus, we want a Fourier Series representation of the Fourier Transform $\hat{f}(s)$.

Specifically, we have from the complex Fourier representation 2.3.3

$$\hat{f}(s) = \frac{1}{\sqrt{2\pi}} \sum_{k=-\infty}^{\infty} c_k e^{iks} \tag{5.2}$$

where

$$c_k = \frac{1}{\sqrt{2\pi}} \langle \hat{f}(s), e^{iks} \rangle. \tag{5.3}$$

Recall that this representation is only valid on $[-\pi, \pi]$. Since $\hat{f}(s) = 0$ for $|s| > \pi$, we can consider working on

$$\hat{f}(s) = \hat{f}(s)\chi(s)$$

where

$$\chi_\pi(s) = \begin{cases} 1 \text{ if } |s| < \pi \\ 0 \text{ if } |s| > \pi \end{cases}. \tag{5.4}$$

The important point is that the representation in 5.2 is only valid in $[-\pi, \pi]$, but we can validly write

$$\hat{f}(s) = \left(\frac{1}{\sqrt{2\pi}} \sum_{k=-\infty}^{\infty} c_k e^{iks} \right) \chi_\pi(s)$$

and this representation is valid for all $s \in (-\infty, \infty)$. We can multiply the $\chi_\pi(s)$ through the sum, and we have

$$\hat{f}(s) = \frac{1}{\sqrt{2\pi}} \sum_{k=-\infty}^{\infty} c_k \chi_\pi(s) e^{iks}.$$

The key is to calculate $f(t)$ via the Inverse Fourier Transform of this Fourier Series representation,

$$f(t) = \frac{1}{\sqrt{2\pi}} \int \hat{f}(s) e^{-ist} ds$$

$$= \frac{1}{\sqrt{2\pi}} \int \left(\frac{1}{\sqrt{2\pi}} \sum_{k=-\infty}^{\infty} c_k \chi_\pi(s) e^{iks} \right) e^{-ist} ds \qquad (5.5)$$

There is no problem exchanging the integral and the sum, which gives us

$$f(t) = \frac{1}{2\pi} \sum_{k=-\infty}^{\infty} c_k \int \chi_\pi(s) e^{iks} e^{-ist} ds$$

$$= \frac{1}{2\pi} \sum_{k=-\infty}^{\infty} c_k \int \chi_\pi(s) e^{-is(t-k)} ds. \qquad (5.6)$$

We have two tasks to finish (1) Calculate the transform $\int \chi_\pi(s) e^{-is(t-k)} ds$, and (2) Find out what the coefficients c_k are.

Beginning with (1) we notice that $\int \chi_\pi(s) e^{-is(t-k)}$ is merely a translation of $\int \chi_\pi(s) e^{-ist} ds$. So we calculate

$$\int_{-\infty}^{\infty} \chi_\pi(s) e^{-ist} ds = \int_{-\pi}^{\pi} 1 e^{-ist} ds = \frac{1}{-it} e^{-ist} |_{-\pi}^{\pi},$$

$$= \frac{1}{-it} \left(e^{-i\pi t} - e^{i\pi t} \right)$$

$$= \frac{1}{-it} \left(-2i \sin(\pi t) \right)$$

$$= \frac{2 \sin(\pi t)}{t}.$$

Plugging this directly into 5.6, we get

$$f(t) = \frac{1}{2\pi} \sum_{k=-\infty}^{\infty} c_k \int \chi_\pi(s) e^{-is(t-k)} ds$$

$$= \sum_{k=-\infty}^{\infty} c_k \frac{\sin(\pi(t-k))}{\pi(t-k)}. \qquad (5.7)$$

We now have to evaluate the c_k's to finish the proof. But recall from 5.3 that

$$c_k = \frac{1}{\sqrt{2\pi}} \langle \hat{f}(s), e^{iks} \rangle = \frac{1}{\sqrt{2\pi}} \int \hat{f}(s) e^{-iks} ds \qquad (5.8)$$

$$= \mathcal{F}^{-1}(\hat{f})(k) = f(k), \qquad (5.9)$$

by the definition of the complex inner product, and the inverse Fourier Transform. Therefore, we have our result

$$f(t) = \sum_{k=-\infty}^{\infty} f(k) \frac{\sin(\pi(t-k))}{\pi(t-k)}.$$

$\square$

In general, the function $\sin(x)/x$ which from the Shannon Sampling Theorem is referred to as the **sinc** function. Note that it formally does not exist at zero, but it is defined by its limit which is one at zero. That is also consistent with the inverse Fourier Transform.

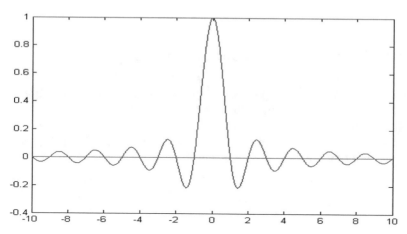

Figure 5.1: Sinc function $\sin(\pi t)/(\pi t)$ from the above theorem is plotted above. Note that it is 1 at 0 and 0 at all of the integers k. While the maxima and minima of this function decrease as t increases, we would like them to decrease faster.

Theorem 5.1.2 *Let $f(t)$ be a function in $L^2(\mathbb{R})$, and let $\hat{f}(s)$ be its Fourier Transform. If $\hat{f}(s) = 0$ for $|s| > \Omega$, or an Ω-band limited functions, then we can represent $f(t)$ exactly by the values of $f(t)$ at the integers $f(k/\Omega)$, and the representation is given by*

$$f(t) = \sum_k f\left(\frac{k\pi}{\Omega}\right) \frac{\sin(\Omega(t-k))}{\Omega(t-k)} \qquad (5.10)$$

where $k = -\infty \cdots - 2, -1, 0, 1, 2, \ldots \infty$. Thus, you can exactly represent any Ω-band limited function by an appropriate collection of samples.

The relationship between Ω and how many samples are needed is direct. If a function g has twice the bandwidth of a function f, then g will require twice as many samples to represent it exactly.

Interpolation

We would like to formally define interpolation.

Definition 5.1.2 (Interpolating functions) *Given data $f(t_k)$ for a number data points t_k, find a continuous function g which obeys the relationship $f(t_k) = g(t_k)$. Any such function g is called an interpolating function for f at the points t_k.*

The Shannon Theorem is much more than a way to construct an interpolating function. The Shannon Theorem is an exact interpolation of $f(t)$, which is valid at all points t given the constraints of the theorem, not just at the interpolation points t_k which are the integers k in the Theorem stated above.

5.1.1   Recalling Polynomial Interpolation

We should contrast the Shannon Sampling Theorem against a fairly standard polynomial interpolation methods, specifically Lagrange interpolation. Lagrange interpolation involves constructing basis polynomials $L_k(t)$ which are 1 at one particular point t_k and zero for all other points t_j, $k \neq j$. This is done simply with the following formula

$$L_k(t) = \prod_{\substack{j = 0 \\ j \neq k}}^{j=n} \frac{t - t_j}{t_k - t_j}.$$

Note that $L_k(t_j) = \delta(i,j)$.

   This yields a very specific interpolation function

$$p_n(t) = \sum f(t_k)L_k(t). \tag{5.11}$$

Note that a) $p_n(t)$ is an n'th degree polynomial, b) $p_n(t_k) = f(t_k)$ so $p_n(t)$ is a interpolating function $f(t)$.

   The Shannon Theorem guarantees that the representation will be valid for π-band limited functions. The polynomial interpolation $p_n(t)$ in 5.11 is also exact, for a different class of functions. It is exact for functions which are n'th degree polynomials.

5.1.2 Shannon's Theorem: Advantages and Disadvantages

The Shannon interpolation is exact for functions whose Fourier Transform is zero outside a certain region. The Lagrange interpolation is exact for polynomials of degree less than or equal to n, when $n + 1$ points are used. With realistic data, neither of these conditions are usually met. In realistic situations where the data comes from an unknown source, there is no right answer. Experience and preference in a given field generally guide the user. In many applications, the Shannon Theorem is used more extensively.

Interpolation with polynomials can be done for large data sets. Generally low-order polynomials are used on a small portion of the data, with each interpolation utilizing only a few of the points. This is referred to a piecewise polynomial interpolation. While there are some interesting applications for this, we will stick to Fourier approaches in this book.

There are a number of **advantages** to using the Shannon Theorem rather than polynomial interpolation. For clarity, we will list them below.

- **Decay:** Comparing Figure 5.1 to Figure 5.2, it is rather obvious that the base interpolating function used in the Shannon Theorem, i.e., the sinc

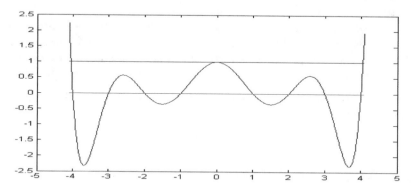

Figure 5.2: We have plotted a Lagrange polynomial which is 1 at zero and 0 at the other integers from -4 to 4. Note that it has very significant ripples away from zero and is not well localized. This is even worse if we have plotted beyond 5, where it goes quickly to ∞. Thus, the interpolating function from a Lagrange polynomial on a large number of points is unacceptable in general. It is informative to compare this to the interpolating function from Shannon's Theorem, shown in Figure 5.1. Note the large maxima at ± 3.7 or so, which are even larger than 1. This implies that the value $f(0$ will strongly affect values away from zero.

function, is better behaved than the Lagrange polynomials. How do we justify this claim? To begin with, the sinc function does decay to zero as $t \rightarrow \infty$. Secondly, with the exception of the sinusoidal variation, the maxima and minima of the sinc function uniformly decay as the distance from the center, or interpolation, point increases. With the Lagrange polynomial, as illustrated in Figure 5.2, neither of these conditions are satisfied. The Lagrange polynomial approaches infinity with the appropriate polynomial speed, as soon as the last of the point of interpolation is outside of the region where the data points are located. In addition, notice that the maxima and minima between interpolation points actually increase.

- **Uniformity:** The interpolation in Shannon's Theorem utilizes the transform of one function. That is certainly not true with polynomial interpolation. Indeed all of the polynomials in Lagrange interpolation are different. This uniformity also allows for rapid calculation of the interpolated function. We will discuss this as we explore the computational methods for applying Shannon's Theorem.

While Shannon's Theorem is very nice, there are also a number of **disadvantages** to the theorem.

- **Slow Decay:** While the sinc function does decay, it decays very slowly, i.e., $O(1/t)$. As a result, the direct computation using the Shannon Theorem will require contributions from nearly all of the data points to accurately represent the interpolated function. It is desirable to replace the sinc function with a function which decays rapidly. Then one would only have to know a few values around an interpolation point in order to find the interpolated value at that point.
- **Uniformity:** If the sampled data is not uniform, then the Shannon Theorem is not applicable. At the least, a lot of work would be required to use "Shannon type" interpolation. This is not a problem with polynomial interpolation and is therefore a strength of polynomial interpolation.

We will now investigate a general approach to sampling, which will allow us to minimize some of the problems or disadvantages above.

5.2 Advanced Interpolation Theory

5.2.1 Basic Ideas

We will now start to formalize our ideas on how to improve upon the basic Shannon Sampling Theorem. While we noted two problems in the last section, slow decay and uniformity, we will concentrate on slow decay. We want to concentrate on this problem for a number of reasons. Let us return to the

Shannon formula 5.10, where we express $f(t)$ as a sum of the samples $f(k)$ times the interpolating function $sinc(x)$, or

$$f(t) = \sum_{k=-\infty}^{\infty} f(k)\frac{\sin(\pi(t-k))}{\pi(t-k)}. \tag{5.12}$$

This is a beautiful formula, but we must now try to understand the difficulties in applying it. If we want to employ this formula to evaluate $f(t)$ at a point t then we need to ask a number of questions. First of all, since we normally won't be able to evaluate the infinite sum, we must figure out how many terms we need. Namely, given a fixed error bar ϵ is there a number N such that if we compute the truncated sum,

$$f(t) \approx \sum_{k=-N}^{N} f(k)\frac{\sin(\pi(t-k))}{\pi(t-k)}, \tag{5.13}$$

we will be able prove to that the error between 5.23 and 5.14 is less than ϵ. The answer is yes, but it is not a strong yes. Specifically, considering the difference between $f(t)$ as represented in 5.23 and subtracting the approximation 5.14, we get

$$f(t) - \sum_{k=-N}^{N} f(k)\frac{\sin(\pi(t-k))}{\pi(t-k)} = \sum_{|k|>N} f(k)\frac{\sin(\pi(t-k))}{\pi(t-k)}. \tag{5.14}$$

Thus, we must bound the inner product on the right, which is bounded as best as possible by the Cauchy–Schwartz inequality or

$$\left| \sum_{|k|>N} f(k)\frac{\sin(\pi(t-k))}{\pi(t-k)} \right| \leq \|\{f(k)\}\|_2 \left\| \left\{ \frac{\sin(\pi(t-k))}{\pi(t-k)} \right\}_{|k|>N} \right\|_2. \tag{5.15}$$

We cannot control the term $\|f(k)\|_2$, which is equivalent, by the Fourier isometry to $\|f(t)\|$. This is fine. If the function we are interpolating is larger, we will need more terms to control the error. We must control the right-hand side of 5.15. For simplicity, we will assume that $|t| < 1$, and since the error term is symmetric, we have

$$\left\| \left\{ \frac{\sin(\pi(t-k))}{\pi(t-k)} \right\}_{|k|>N} \right\|_2^2 = \sum_{k=N+1}^{\infty} \left| \frac{\sin(\pi(t-k))}{\pi(t-k)} \right|^2$$

$$\leq 2 \sum_{k=N-1}^{\infty} \left| \frac{\sin(\pi(t-k))}{\pi(t-k)} \right|^2$$

$$\leq 2 \sum_{k=N-1}^{\infty} \left| \frac{1}{\pi(t-k)} \right|^2$$

$$= \frac{2}{\pi} \sum_{k=N-1}^{\infty} \frac{1}{k^2}. \tag{5.16}$$

Now by basic calculus, we can bound this by the integral which gives us

$$\left\| \left\{ \frac{\sin(\pi(t-k))}{\pi(t-k)} \right\}_{|k|>N} \right\|_2^2 \leq \frac{2}{\pi} \sum_{k=N-1}^{\infty} \frac{1}{k^2}$$

$$= \frac{2}{\pi} \int_{N-2}^{\infty} \frac{1}{t^2} dt = \frac{2}{\pi(N-2)}. \tag{5.17}$$

Thus, we have that the error in 5.15 is bounded by the inequality

$$\left| \sum_{|k|>N} f(k) \frac{\sin(\pi(t-k))}{\pi(t-k)} \right| \leq \|f(t)\|_2 \sqrt{\frac{2}{\pi(N-2)}}$$

$$= O\left(\frac{1}{\sqrt{N}}\right) \|f(t)\|_2. \tag{5.18}$$

What does this mean?

When we develop error bounds, the mathematics is fine, but one has to ask after the fact, "What does this mean?". Let us try to put that in perspective. Let us suppose that you want the error 5.18 to be bounded by ϵ, then you need

$$\left| \sum_{|k|>N} f(k) \frac{\sin(\pi(t-k))}{\pi(t-k)} \right| \leq O\left(\frac{1}{\sqrt{N}}\right) \|f(t)\|_2 < \epsilon \|f(t)\|_2. \tag{5.19}$$

This means that

$$\epsilon = O\left(\frac{1}{\sqrt{N}}\right)$$

or that $N\epsilon^2 = O(1)$ or $N = O(\frac{1}{\epsilon^2})$. To understand what this means, let's put real numbers in for ϵ. If $\epsilon = 1/100$, or we want $1/100$ as our maximum relative error, then we need N to be $100^2 = 10,000$. If you want $1/1000$, you need 1000^2 coefficients. So for a 1% error in the region about zero, we need to collect and evaluate 10000 samples. This points out the weakness of the Shannon Sampling Theorem at its worst.

Moreover, this means that the values of 10,000 different coefficients would be needed to take care of this one evaluation point. If there is a mistake in any one of them, then the whole interpolation will be affected.

The reason why all of these coefficients are necessary, and all values are affected by all others, is that the sinc function decays so slowly. Therefore, we want to address this and find a remedy. To this end, let us introduce a couple of concepts.

Definition 5.2.1 (Local r-Support) *A function $f(t) \in L^2(\mathbb{R})$ is said to be locally r-supported about t_0 if there is a number r such that $f(t) = 0$ for $|t - t_0| > r$. We refer to a function which is r-supported for any finite r as a compactly supported function.*

Note that the sinc function is not locally r-supported at all. Fourier Analysis, however, is oftentimes associated with controlling errors. Thus, we are interested not only in functions which are zero outside an interval, but also in functions which are nearly zero outside a finite region. To formalize this idea, we introduce the idea following definition.

Definition 5.2.2 (Local r(ϵ)-Support) *A function $f(t) \in L^2(\mathbb{R})$ is said to be locally r(ϵ)-supported about t_0 if there is a number r such that $|f(t)| < \epsilon$ for $|t - t_0| > r$.*

The sinc function is r(ϵ)-supported, but the relationship between r and epsilon, from above, is $r = O(1/\epsilon)$. Thus for ϵ small, r must be very large. We want to alter the way we are choosing things so that the interpolation function which we are using is more tightly supported.

5.2.2   Oversampling and Adaptive Windows

Let us go back to the Shannon Sampling Theorem, and remember why we had to use the sinc function, and think about how to get around it. The basic requirement for the Theorem was that $\hat{f}(s) = 0$ for $|s| > \pi$. We then represented the function $\hat{f}(s)$ in the standard Fourier Series on $[-\pi, \pi]$ or

$$\hat{f}(s) = \sum_k c_k e^{iks},$$

which is valid on $[-\pi, \pi]$, but is periodic, and therefore not a valid representation for $f(t)$ beyond $[-\pi, \pi]$. We extended this representation by multiplying by $\chi_\pi(s)$ which is $1 \in [-\pi, \pi]$ and 0 elsewhere, to get a representation

$$\hat{f}(s) = \sum_k c_k e^{iks} \chi_\pi(s),$$

which is valid on $(-\infty, \infty)$. The Inverse Fourier Transform of this series then yielded the Shannon Sampling Theorem.

The problem with this is the function $\chi_\pi(s)$ which is $1 \in [-\pi, \pi]$ and 0 elsewhere. This is a discontinuous function. As we learned earlier, a discontinuous function will have a Fourier Transform, or Inverse Fourier Transform, which does not decay quickly. Thus, we have the problems above. But in this setting, we needed $\chi_\pi(s)$ to be $1 \in [-\pi, \pi]$ and 0 elsewhere, since the representation was 2π periodic.

Oversampling

We will now consider altering our Fourier Series for $\hat{f}(s)$ so that we can represent it not only on $[-\pi, \pi]$ but also on $[-2\pi, 2\pi]$. Since $\hat{f}(s) = 0$ for $|s| > \pi$ it follows that the representation will converge to 0 for $\pi < |t| < 2\pi$. The representation for $\hat{f}(s)$ on $[-2\pi, 2\pi]$ is given by

$$\hat{f}(s) = \sum_k c_k e^{iks/2}, \tag{5.20}$$

with the c_k's given by the appropriate formulas in Theorem 2.3.3. We can multiply by $\chi_{2\pi(s)}$ and we get a representation

$$\hat{f}(s) = \sum_k c_k e^{iks/2}\chi_{2\pi}(s), \tag{5.21}$$

which is valid on $(-\infty, \infty)$. The Inverse Fourier Transform gives us an additional Shannon Sampling representation

$$f(t) = \sum_k f\left(\frac{k}{2}\right) \frac{\sin(\pi/2(t-k))}{\pi/2(t-k)}. \tag{5.22}$$

Comparing this to the original Shannon Theorem

$$f(t) = \sum f(k) \frac{\sin(\pi(t-k))}{\pi(t-k)}, \tag{5.23}$$

we see that we are using twice as many samples in 5.26 as in 5.23, thus we call this an **oversampled** representation. We have more samples than necessary. We will use this shortly.

The Shannon representation 5.26, however, does not have many advantages to that in 5.23. It requires twice the number of samples and gives the same result. Moreover, the r(ϵ) supports of the functions are very equivalent.

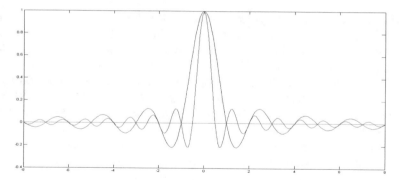

Figure 5.3: We illustrate two sinc functions above. The first is the Inverse Fourier Transform of a χ_π, and the second is the Inverse Fourier Transform of $\chi_{2\pi}$. Note that the first sinc function is zero at all of the integers k and the second sinc function is zero at the half integers $k/2$. The second function can be used to interpolate using the values at $k/2$, while the first requires the values at k. Note that the Shannon Sampling Theorem dictates the sampling rate.

Let us back up and reexamine the representation of the Fourier Transform in 5.25. This representation is valid from $[-2\pi, 2\pi]$ so we have a new freedom. In the past, we had to use $\chi_\pi(s)$ which is 1 on $[-\pi, \pi]$ and zero elsewhere to get a representation which is valid on $(-\infty, \infty)$. We can now choose any window function which is 1 on $[-\pi, \pi]$ and zero outside of $[-2\pi, 2\pi]$ since the Fourier representation 5.20 is 0 between $\pm\pi$ and $\pm 2\pi$. Thus, we don't need to have a discontinuity in our window function.

5.2.3 Interpolating Window Functions

Restating the story, since the representation in 5.20 is zero for $\pi < |t| < 2\pi$, we can choose any window function $w(t)$ such that

$$w(s) = \begin{cases} 1 \text{ if } |s| < \pi \\ 0 \text{ if } |s| > 2\pi \end{cases}, \tag{5.24}$$

and we will have the representation

$$\hat{f}(s) = \sum_k c_k e^{iks/2} w(s), \tag{5.25}$$

which will be valid from $(-\infty, \infty)$. We see from above that there are no conditions on $w(s)$ from π to 2π. This is where oversampling gives us the ability to build a better system. The question is "What do we do with the function from π to 2π?"

If we are to have an interpolation function which has optimal decay, namely which is r(ϵ) with $r(\epsilon)$ as small as possible. To do this, recall that the window $w(t)$ will have to be smooth. If there is a discontinuity in the window function, the Inverse Fourier Transform of the function must decay slowly. Recall that the Fourier Transform of $\chi_\pi(s)$ is the sinc function and since there is a discontinuity in $\chi_\pi(s)$, the sinc function decays slowly, resulting in the bad results of the last section. Namely, we needed approximately 10000 coefficients to get an accuracy of .01.

With oversampling, we have the opportunity to eliminate the discontinuity in the window function. We can have a smooth function which transfers from 1 at π to 0 at 2π smoothly. Taking the Inverse Fourier Transform of 5.25, we get

$$f(t) = \sum_k f\left(\frac{k}{2}\right) \hat{w}(t - k/2). \tag{5.26}$$

Let us reiterate the rules for a proper window function. We will refer to these rules as the **window criteria**.

Definition 5.2.3 (Window Criteria) *A function $w(t)$ is said to satisfy the window criterion on the embedded intervals $[-\pi, \pi]$ and $[-2\pi, 2\pi]$, if it satisfies the following conditions.*

- *The window $w(t)$ must be 1 from $[-\pi, \pi]$.*
- *The window $w(t)$ must be 0 for $|s| > 2\pi$.*
- *The window $w(t)$ must make a continuous transition from π to 2π and from $-\pi$ to -2π. In other words, the function must be continuously differentiable on $(-\infty, \infty)$.*

We similarly define the window criterion on any set of embedded intervals $[a1, b1]$ and $[a, b]$, where $a1 < a < b < b1$.

We would like the window to also be continuously differentiable, and perhaps have higher derivatives which are continuous. The question comes up, "How continuous, in the third criterion must the window be?" That is a good question. Why does it need to be continuous? Because if it is, then the interpolating function will be "nicely" r(ϵ)-supported. How many times continuous is not clear. Let's explore it.

The Raised Cosine

We will begin with one of the standard interpolating windows. The goal is to transition from 1 to 0 in a very continuous manner. One way would be standard linear interpolation, but that would leave a discontinuity in the derivative at the transition point. We want a transition which is continuous and has continuous derivatives. One standard way to do this is with what is called the **raised cosine window**.

Let us start out with the basic function

$$rc(t) = \frac{1}{2}(\cos(t) + 1).$$

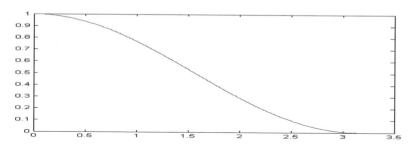

Figure 5.4   Raised cosine is plotted above. Notice the smooth transition from 1 to 0.

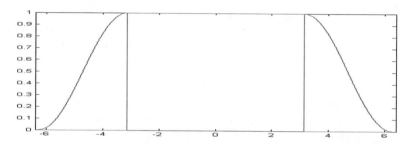

Figure 5.5:   An interpolating window constructed from the raised cosine, according to 5.27, is plotted above. Notice the smooth transition from 1 to 0. This is contrasted against the standard $\chi_\pi(s)$ function with the direct transition from 1 to 0.

Notice that analytically, $rc'(t) = 1/2\sin(t)$, so $rc'(0) = 0$, and $rc'(\pi) = 0$. We want to pair this with a characteristic function χx.

What we want to do it create window function for $[-2\pi, 2\pi]$ which satisfies the criterion 5.2.3. We can do this by using

$$w_{rc}(s) = \begin{cases} rc(-(s - \pi)) & \text{if} & -2\pi < s < \pi \\ 1 & \text{if} & |s| < \pi \\ rc(s - \pi) & \text{if} & \pi < |s| < 2\pi \\ 0 & \text{if} & |s| > 2\pi \end{cases}, \qquad (5.27)$$

Notice that $w_{rc}(s)$ is continuous and differentiable at $\pm\pi$, and $\pm 2\pi$. By this, we note that the left-hand limit and left-hand derivative limits are the same as the right-hand limits and derivative limits at those points. However, $rc''(t) = -1/2\cos(t)$, so $rc''(0) = -1/2$ and therefore the window $w_{rc}(t)$ does not have continuous second derivatives. Thus, the Fourier Transform of w_{rc} cannot decay too quickly, since its second derivative does not exist. Let us test this numerically, to see if it decays rapidly enough for practical use.

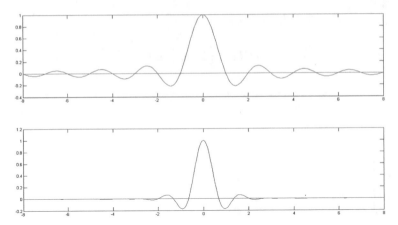

Figure 5.6: Fourier Transforms of the smoothly windowed raised cosine window and the discontinuous function χ_π or the sinc function are shown above. The sinc function is at the top, and the smooth raised cosine window is at the bottom. Notice the smooth transition from 1 to 0 for the raised cosine window resulted in a function which was essentially $r(\epsilon)$-supported, while the Fourier Transform of the χ_π function continues to ring on. Specifically, if we choose ϵ to be .01, then the raised cosine window is r(.01) supported on $[-3,3]$. The function χ_π has r(.01) supported on $[-100:100]$.

The point of utilizing smooth windows is to localize, in and $r(\epsilon)$ type of way, the interpolation function $w_{rc}(t)$. To illustrate this, we give you the Inverse Fourier Transforms of the functions $\chi_\pi(t)$ and $w_{rc}(t)$ in Figure 5.6.

From the illustration in Figure 5.6 and the numerical verification of its $r(\epsilon)$-support, we see that the raised cosine generated window does a relatively good job of localizing the interpolation function. There are an infinite number of interpolating windows. A standard internet search will give a variety of approaches. The question of which is best can only be determined by the application. We leave a couple of additional examples in the exercises.

5.2.4 Approximate Window functions:

In the last section, we looked at one of many possible window functions which allow one to have the exact interpolation formula 5.26

$$f(t) = \sum_k f\left(\frac{k}{2}\right)\hat{w}(t-k/2). \tag{5.28}$$

Any function $w(t)$ which satisfies the **window criteria** will admit this type of representation. Oftentimes, however, we do not need equality in the above equation, but will happily settle for approximate equality. Thus, we want

$$f(t) \approx \sum_k f\left(\frac{k}{2}\right) \hat{w}(t - k/2). \tag{5.29}$$

and of course one would like to be control the error or have $\| f(t) - \sum_k f\left(\frac{k}{2}\right) \hat{w}(t - k/2) \| < \epsilon$, where ϵ is chosen according to the application. For instance, when storing medical images, it is generally unacceptable to have ϵ be significant at all, since the examination of those images is used to detect cancer or other serious questions. On the other hand, if one can supply more TV channels, and the quality is not immediately noticeable by the user, then this is very useful and very valuable.

Additionally, since we are approximating these functions we would like to consider the approximation to the Shannon Theorem

$$f(t) \approx \sum_k f(k)\hat{w}(t - k). \tag{5.30}$$

It is then assumed that the $r(\epsilon)$-support of $\hat{f}(s)$ is contained within $[-\pi, \pi]$.

There are other reasons for utilizing approximate window functions. As we stated to begin this chapter, the criterion for utilizing the Shannon Sampling Theorem is almost never satisfied, but only approximately satisfied. Therefore, the equality in the Shannon expansions is generally not equality but approximate equality. Therefore if we allow ourselves to choose a window which is "nicer" in some ways, but does not satisfy the window criterion, this may not be all that noticeable.

The question then is, "What criterion do we want for approximate windows?". This is not well defined, but we will try to give an answer, which is certainly not the only answer.

Definition 5.2.4 (Approximate Window Criteria) *Let a function* $f(t) \in L^2(\mathbb{R})$ *and for some* $\epsilon > 0$ *suppose that the Fourier Transform* $\hat{f}(s)$ *is* $r(\epsilon)$-*supported. Then we would say a function* $w(s)$ *is said to satisfy the approximate window criterion if it satisfies the following conditions.*

1. *The window* $w(s)$ *must be 1 at 0.*

2. *The window* $w(s)$ *must be positive.*

3. *The window* $w(s)$ *must be continuous and hopefully continuously differentiable.*

4. *The* $r(\epsilon)$-*support of the function* $w(s)$ *should contain or be essentially the same as the* $r(\epsilon)$-*support of the Fourier Transform* $\hat{f}(s)$.

These criterion are really good "rules of thumb," and all of them are fairly flexible. The first criterion just means that the average value of f should not change. The second criterion is generally used, because while weighting the

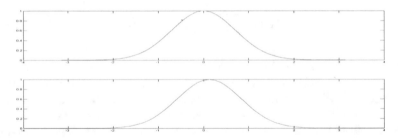

Figure 5.7: Gaussian function $\exp(-t^2)$ is illustrated above and its Fourier Transform, which was calculated numerically, is illustrated below. Note that they are nearly identical, with a minor adjustment to the scaling needed. Note also how well localized the Gaussian is. If $\hat{f}(s)$ were only supported in $|s| < \pi$, then an approximate Shannon expansion such as 5.30, with this very well localized function instead of the broadly supported sinc function.

coefficients, or Fourier Transform of the function is normal, giving negative weight is not unless there is a specific application in mind. The third criterion should generate a "nicely" $r(\epsilon)$-supported function. The fourth assures that the frequency content of f, while changed, is still essentially present in the interpolated approximation.

The Gaussian et. al.

There are a number of possibilities for functions which meet these criteria. The most obvious is the Gaussian $\exp(-as^2)$. It obviously satisfies criterion (1) and (2). It is infinitely continuous and differentiable, so it satisfies (3) very well. Criterion 4) can be satisfied by choosing a sufficiently small to include the $r(\epsilon)$-support of the Fourier Transform of $\hat{f}$ in the essential support of the function $w(s)$. The Gaussian is oftentimes a preferred function, because it satisfies the uncertainly principle, meaning that its support in time is a short as possible given its bandwidth in frequency. We illustrate this function in Figure 5.7.

Notice that the exponential in Figure 5.7 is essentially supported inside of $[\pi, \pi]$. If $\hat{f}(s)$ is also similarly supported then the approximate expansion 5.30 can be used. Since the Fourier Transform of this window has essential support which is only slightly larger than $[-2, 2]$, only at most 5 terms from 5.30 would be necessary at each point t. Specifically, for any given t, there can only be at most terms $t - k$ which are between $[-2, 2]$, or perhaps $[-2.5, 2.5]$.

One could also use $1/(1 + (at)^2)$. This dies more slowly than the Gaussian, but may be of value.

Once again there are an infinite number of windows. Many of them have been attributed to one or another researcher who first used them. The most important part of choosing a window is to understand your goals and to construct the window which works for you.

5.2.5   Problems and Exercises:

1. Consider a window made with a raised cosine and one constructed approximately from two half's of a Gaussian. With the same transition time from 1 to 0, which one seems better?

2. Consider the function $f(t) = exp(-1/t^2)$ for $t > 0$ and 0. How many derivatives does it have at 0.

3. Consider the transition function $f(t) = 1/2(1 + \cos(t))$ for $0 \leq t \leq \pi$, the raised cosine. Now consider $f^2(t)$. How many derivatives will $f(t)$ supply?

5.3   Aliasing

Thus far in this chapter, we have talked about exact sampling, via the standard Shannon Sampling Theorem. We have talked about oversampling and the freedom that this provides for choosing a different window function $w(t)$ which will be "nicer" interpolation function. We have also talked approximate windowing and the freedom that this allows for a "nice" interpolation.

We have not, until now, explored the consequences of under sampling. Specifically, we have noted that if a function does not exactly satisfy the conditions of the Shannon Sampling Theorem 5.1.1, then the equality will not hold. Moreover we have mentioned that functions almost always do not satisfy the conditions of 5.1.1. We will now quantify the errors which occur when a function is not appropriately band limited, or when $\hat{f}(s) \neq 0$ for all $|s| > \pi$. If one uses the Shannon Theorem, or any similar theorem, and the conditions on it Fourier Transform are not adequately satisfied, then we call this an under sampling condition.

Under sampling is quite literally the attempt to represent a function with too few samples. Any such attempt is doomed to failure, sometimes a failure which is small, and sometimes catastrophic failure. We need to understand which of these conditions will be present.

The Wagon Wheels

A classic example of under sampling is oftentimes seen in old Western cowboy movies. There is often a stage coach, which is running away at high speed from robbers or Native Americans. The under sampling is seen when the spoked wheels on the stage coach appear to turn slowly either forward or backward. This is because the old-fashioned cameras were not sufficiently fast to capture the speed of the wheels. The speed of sampling was such that one spoke would move nearly to the location of another spoke, during the time the camera took to record the different slides of the movie. Thus, it appears that this first spoke actually moved backward to the location of the second spoke.

Although this example is amusing, it is not an example of catastrophic sampling, where we believe that something is true just because we did not sample well. We all know that the wagon wheel is turning forward, so we disregard what the picture looks like. The result can be much worse when we are trying to discover things which we don't understand.

The Common Misconception of Aliasing

One very common misconception when we consider aliasing is that aliasing errors are only a high frequency problem. To be more specific, since the Shannon Sampling Theorem represents functions which have only low-frequency content or $\hat{f}(s) = 0$ for $|s| > \pi$, then we would like to believe that the using the Shannon representation will give us a low-frequency approximation to the actual function, even if this condition is not met. This is not true and understanding why is fundamental to understanding aliasing. We will first define the low and high frequency approximations to a function $f(t)$.

Definition 5.3.1 (Bandpass Approximations) *Given a function $f(t) \in L^2(\mathbb{R})$ and a pass band parameter α, we define the low pass and high pass approximations to $f(t)$ by*

$$f_{low}(t) = \mathcal{F}^{-1}\left(\hat{f}(s)\chi_{[-\alpha,\alpha]}(t)\right),$$

and

$$f_{high}(t) = \mathcal{F}^{-1}\left(\hat{f}(s)(1 - \chi_{[-\alpha,\alpha]}(t))\right),$$

where $\chi(t)$ was defined in 2.59. Note that $f = f_{low} + f_{high}$. We will refer to f_{low} as the low pass approximation and f_{high} as the high pass approximation.

Let us assume for now that $\alpha = \pi$ and consider using the Shannon Sampling representation 5.10. If $f_{high}(t) = 0$ for all t, then the Shannon representation is valid, but more realistically, we generally don't know this. So we have the Shannon representation with assumed bandwidth π, which we will denote by $S_\pi(f(t))$ and which is given by

$$S_\pi(f(t)) = \sum_k f(k)\frac{\sin(\pi(t-k))}{\pi(t-k)}$$

$$= \sum_k f_{low}(k)\frac{\sin(\pi(t-k))}{\pi(t-k)} + \sum_k f_{high}(k)\frac{\sin(\pi(t-k))}{\pi(t-k)}. \quad (5.31)$$

Note that the proper Shannon representation for $f_{low}(t)$ with $\alpha = \pi$ is the representation on the left in 5.31. Therefore, we have

$$S_\pi(f(t)) = f_{low}(t) + \sum_k f_{high}(k) \frac{\sin(\pi(t-k))}{\pi(t-k)}$$
$$= f_{low}(t) + A(t),$$

where we have defined an aliasing function above. At first glance one might be hopeful that the Shannon approximation is primarily $f_{low}(t)$ as we see on the right of 5.32. The second term, however, is not generally zero. Moreover, the second term is a low-frequency term. Even though the coefficients $f_{high}(k)$ come from the high pass version of f, they are attached to the low-frequency terms

$$\frac{\sin(\pi(t-k))}{\pi(t-k)}.$$

Thus, the Fourier Transform of $S_\pi(f(t))$ is given by

$$\mathcal{F}(S_\pi(f(t))) = \hat{f}(s)\chi_\pi(s) + \sum_k f_{high}(k)e^{iks}\chi_\pi(s),$$
$$= \hat{f}(s)\chi_\pi(s) + \hat{A}(s), \tag{5.32}$$

where we are designating calling $\hat{A}(s)$ the aliasing term in frequency, and $A(t)$ the corresponding aliasing term in time. Thus, the Shannon representation has two low-frequency terms, one of which is the desirable $f_{low}(t)$, and the other which is aliasing.

If we subtract this from f we get

$$f(t) - S_\pi(f(t)) = (f_{low}(t) + f_{high}(t)) - \left(f_{low}(t) + \sum_k f_{high}(k) \frac{\sin(\pi(t-k))}{\pi(t-k)} \right)$$
$$= f_{high}(t) + A(t).$$

Thus, the difference between an arbitrary function $f(t)$ and the incorrectly used Shannon representation $S(f(t))$ consists of the high pass representation $f_{high}(t)$ which we should anticipate. In addition, however, we have the aliasing term $A(t)$ which is given by

$$A(t) = \sum_k f_{high}(k) \frac{\sin(\pi(t-k))}{\pi(t-k)}.$$

This aliasing term corrupts the low frequencies of the Shannon representation. Thus, under sampling not only ignores the high frequencies, but the high frequency contributions corrupt the low frequency representation.

At this point in time, we would like to be able to bound the aliasing error $A(t)$ by the norm of either $f(t)$, or $f_{high}(t)$. This would require some type of isometry between the coefficients $\{f_{high}(k)\}_k$ and the function $f_{high}(t)$.

Unfortunately, no such isometry exists. One of the exercises is to construct a function $f(t)$ such that $\|f(t)\| = 1$ and $\|A(t)\|$ is arbitrarily large.

5.3.1 Anti-Aliasing Filters

The above discussion raises the question again, "How do we know whether the Shannon representatio, or any of the oversampled representations are valid?". If we are only given the samples, then we really cannot tell. Usually, however, we know something about the source of the data. If it is human speech, then we know its bandwidth. Similarly with light and other phenomena.

There is another method for controlling the error $A(t)$, however. Assuming that we are taking discrete samples from a continuous stream of data, we can control $A(t)$, oftentimes with physical hardware devices such as the antennae, or microphone which we use to detect the data.

A Simple Example: The RLC circuit

The simplest and probably most common example of an anti-aliasing filter is an RLC circuit, which we mentioned in section 4.3.3. The beauty of these circuits is that they are simply constructed with electronic components, a resister (R), an inductor (L), and a capacitor (C). In addition, the circuit is easily described via basic differential equations. The Kirchoff laws can be found in any basic physics or electrical engineering text. They state that the behavior of a signal $y(t)$ in such as system can be described by a simple second order linear differential equation of the form

$$ay''(t) + by'(t) + cy(t) = r(t),$$

where $y(t)$ is the response of the filter and $r(t)$ is the external, forcing signal.
Now utilizing the Fourier Transform of the above, we have

$$as^2\hat{y}(s) + b(is)\hat{y}(s) + c\hat{y}(s) = \hat{r}(s).$$

Solving algebraically, we have that $\hat{y}(s)(as^2 + (ibs) + c) = \hat{r}(s)$ or that

$$\hat{y}(s) = \frac{1}{as^2 + (ibs) + c}\hat{r}(s) = \hat{T}(s)\hat{r}(s),$$

where we refer to $T(t)$ as the transfer function with $\hat{T}(s)$ being given above.
Studying the behavior of $T(t)$ above yields some obvious results. If we let $b = 0$, which is not practical but is informative, then obviously $\hat{T}(s) \to 0$ as $s \to \infty$. This is also true if $b \neq 0$ so any basic electrical circuit tends to be a "low pass" filter, which eliminates the high frequencies. Manipulating the

coefficients a, b, and d through the proper choice of capacitors, resistors, and inductors is the art of analogue filter design. Proper pairing of an analogue filter before sampling will greatly reduce aliasing. In reality, however, aliasing is controlled but never completely eliminated.

5.3.2 Problems and Exercises:

1. **Challenging** Find a function $f(t)$ such that $\|f(t)\| = 1$, and the Aliasing term $A(t)$ has a norm which is N, with N being arbitrarily large.

2. Another problem.

5.4 Numerical Computation of the Shannon Series

In reality, we don't compute the Shannon Series 5.10 directly from the summation form. Instead we utilize the Fourier Transform first, manipulate the frequency content, and then compute the Inverse Fourier Transform of the result to compute the interpolated functions. We also generally don't do this with the Fourier Transform, but rather with the Discrete Fourier Transform (DFT), computed via the Fast Fourier Transform (FFT).

To be specific, the Shannon Series 5.10 assumes that the high frequencies of the function f are all zero. Generally, we are given a finite number of samples $f(k)$, $k = 0, 1, \ldots N - 1$. We generally utilize the 0 to $N - 1$ but 1 to N works, as long as appropriate attention is paid. Now we use the Discrete Fourier Transform to compute $\hat{f} = \{\hat{f}(k)\}$, where $k = 0, 1, \ldots N - 1$.

Let us assume that we have a vector of data denoted by f which has length N. We compute its Discrete Fourier Transform (DFT) and call it $\hat{f}$. As we discussed in Chapter 3, this data is not in the form we would visualize from the continuous Fourier Transform. To reorganize it in more standard form, we shift the data by $N/2$, or compute $S_{N/2}(\hat{f})$ using the the standard MATLAB command fftshift.

The Shannon Theorem specifies that the high frequencies are all zero. Therefore if we want to interpolate to $M > N$ points, we add $(M - N)/2 = q$ zeros to each side of $S_{N/2}(\hat{f})$, maintaining symmetry. Thus, we create the vector

$$S_{M/2}(\hat{f}_M) = [0, 0, 0, \ldots, 0, S_{N/2}(\hat{f}), 0, 0, 0, \ldots, 0], \qquad (5.33)$$

which has $(M - N)/2$ zeros on each side of the Fourier data of the original vector. The final Shannon interpolated vector is then calculated as

$$f_M = F^*(S_{M/2}(S_{M/2}(\hat{f}_M))), \qquad (5.34)$$

where $S_{M/2}(\hat{f}_M)$ is calculated exactly by 5.33. The individual steps in 5.34, must we utilized, since the goal is the calculation of the interpolated data f_M.

5.4.1 Example 1: Exact Interpolation

Let us interpolate a basic function, using the above methods to compute its Shannon interpolation. We will simply use a discrete characteristic function $f_{256} = [0, 0, \ldots, 0, 0, 1, 1, 1, \ldots, 1, 1, 1, 0, 0, \ldots, 0, 0]$ where there are 96 leading zeros, 64 ones, and 96 trailing zeros, for 256 points. This function is illustrated in Figure 5.8.

Now we compute the Fourier Transform of f_{256} and display this and its shifted version $S_{128}(f_{256})$ in Figure 5.9. We would like to display f_{256} on 1024 points. The Shannon Theorem states that all higher frequencies should be zero, so create the expanded vector as above in 5.33. Finally, we invert this as in 5.34, and we display the final results in Figure 5.10.

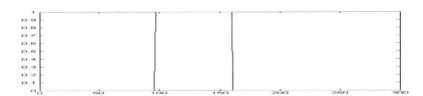

Figure 5.8 Basic function f_{256} is shown above.

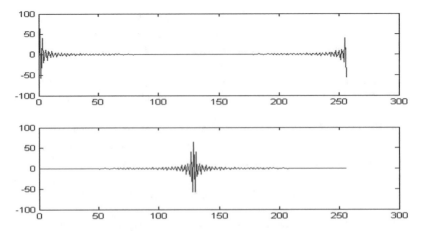

Figure 5.9 DFT of f_{256} and its shifted version $S_{128}(f_{256})$ are shown above.

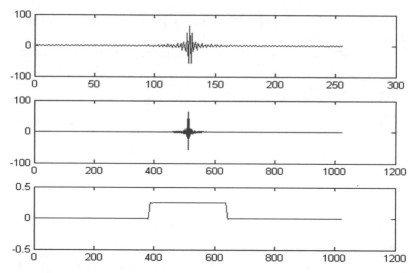

Figure 5.10:   Shifted version of the DFT $S_{128}(\hat{f}_{256})$ are shown at the top. At the center is the same shifted version, but embedded in a 1024 array, or with 384 zeros on either side. Note that the interior is identical but seems compressed because of the increased scale. At the bottom is the Inverse Fourier Transform, or the Shannon interpolated version of the original function. Notice, however, that the interpolated version has a maximum value around 1/4, rather than 1.

There is a problem in the final version. Its value of the interpolated points is 1/4, not 1. This is due to a conversion factor. We interpolated to four times as many points, and the central portion of the DFT remained the same. Therefore, the average went down by a factor of four. This is simply corrected by multiplying by 4. To examine whether this procedure worked well, we consider interior portions of the original and interpolated functions in Figure 5.11.

5.4.2   Example 2: Approximate Interpolation

The example in the last section might mislead us into thinking that Shannon interpolation works very well, without any alterations or adaptations. That example had a) a lot of points to begin with and b) not much structure. We will now consider an example which pushes the limits of Shannon interpolation.

Let us consider the function on 128 points, f_{128} = [zeros(1,32),2*ones (1,16), zeros(1,32),3*ones(1,16),zeros(1,32)], where we have utilized the MAT-LAB conventions of naming zeros and ones. Specifically, zeros(1,32) =

$[0,0,0,\ldots,0,0,0]$ where there are 32 zeros. Similarly for the ones. This function is shown in Figure 5.12.

We notice in Figure 5.12 that there is significant ringing in the pictures.

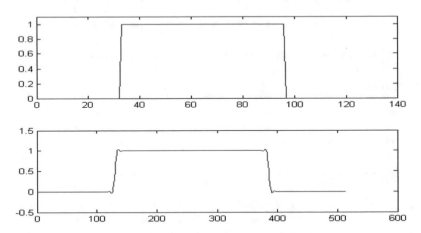

Figure 5.11: This "closeup" of the interpolated functions shows that the interpolation was actually very good. We compensated for the factor of four from the transform, and everything looks relatively good. There is a little bit of ringing at the edges, but not much.

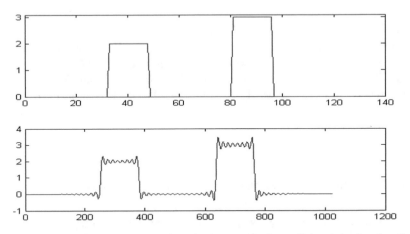

Figure 5.12: Function f_{128} is plotted in the first graph above. Below it is that function, interpolated to 1024 points with the Shannon Sampling Theorem. The method used in example 1 is once again used here. Note the extreme Gibbs ringing which is occuring.

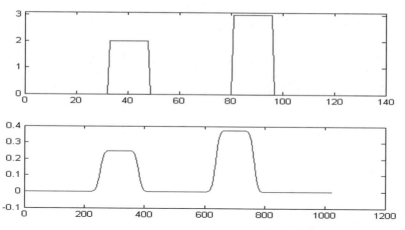

Figure 5.13: Function f_{128} is plotted in the first graph above. Below it is that function, interpolated to 1024 points with the Shannon Sampling Theorem. The method used in Example 1 is once again used here. Note the extreme Gibbs ringing which is occuring.

An appropriate question is whether or not we can fix this, perhaps by using a variant of the Shannon Sampling Theorem. In Figure 5.13, we utilize a non-exact window, a Gaussian window, to somewhat eliminate the ringing.

5.4.3   Problems and Exercises:

1. Replicate Example 1. Specifically use a software package such as MAT-LAB to encode the interpolation using the DFT.

2. Replicate Example 1 with more general choices of functions. Consider functions with discontinuities, fast sinusoidal oscillations, etc.

3. Replicate Example 2. Specifically use a software package such as MAT-LAB to encode the interpolation using the DFT.

4. Replicate Example 2 with more general choices of functions. Consider functions with discontinuities, fast sinusoidal oscillations, etc.

5. Replicate Example 2, but with more general choice of interpolating or approximate interpolating windows. Examples might be the raised cosine, etc.

5.5 Chapter Project:

1. Start out with a signal which is 256 pixels long, such as >> signal = [zeros(1,32),ones(1,64),zeros(1,32),3*ones(1,64),zeros(1,64)]. Use the methods of this section, i.e., the Shannon Theorem implemented through the FFT, to interpolated this to 1024 points.

2. Use a soft window, such as a raised cosine window to cut down on the induced Gibbs ringing in problem 1.

3. Use a window, such as a Gaussian, to cut down on the Gibbs ringing. (You can choose another one if you like).

4. **[Aliasing:]** Consider the following signal = [zeros(1,16),ones(1,8),zeros (1,8), 2*ones(1,32), zeros(1,32),4*ones(1,8),zeros(1,16)]. Subsample this signal to a signal of length 16, or 32. Then try to restore it to a signal of length with the techniques above. Can you get anything close to it back?

5. Consider two signals with t = [1:1024]/1024*2*pi; a) sin(4*t) b) sin(4*t) + sin(32*t). Try to subsample these functions by factors of 16, 32, 64 etc. and then restore them to 1024 points with the techniques above. Can you restore both of them equally well?

Chapter 6
Digital Communications

6.1 Introduction and Basic Terms

Digital communications are at the center of modern life. Cell phones have revolutionized the way we communicate. While they were scarcely available 20 years ago, it is very hard to imagine life without them today. Newer phones and newer TV's have high-resolution digital picture and communication abilities. The details of these systems are highly proprietary, but the fundamentals have already been outlined. They revolve around sampling and interpolation.

We will now explore the basic components of digital communication. Many of these individual components are highly constrained by the Fourier Transform. We are well-equipped at this point in time to address the design issues which arise because of these constraints.

Sample

We illustrate the various parts of a digital communications system in Figure 6.1. We have already covered sampling in Chapter 5. In order to sample correctly, one must sample often enough for the application so that aliasing,

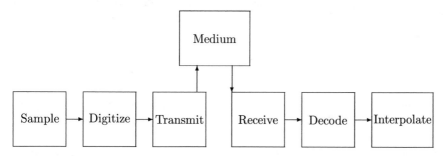

Figure 6.1: A simple schematic for a digital communications system is shown above. The individual components are discussed below.

© Springer Science+Business Media, LLC 2017
T. Olson, *Applied Fourier Analysis*,
https://doi.org/10.1007/978-1-4939-7393-4_6

as described in 5.3, does not become overly destructive. Oftentimes, this is ensured by utilizing an anti-aliasing filter before sampling, as we described in 5.3.1. All of these steps are schematically represented by the first box, Sample, in Figure 6.1.

Digitize

The second portion of the process is the digitization. The idea of digital communications is that only 0's and 1's are transmitted. Thus the samples are broken down into their approximate digital representation. Generally, the range of the samples, speech for example, is finite. This corresponds to the relative volume. Thus there is a maximum and a minimum value that has to be transmitted, or we assume that $-M < f(k) < M$, with M being the joint minimum and maximum.

We need to break this range from maximum to minimum down into finite increments. For single-bit transmission, we use 256 or 2^8 divisions between $-M$ and M, with 0 being assigned to $-M$ and 255 assigned to M. For finer representation, we utilize two-bit enclosing, or 2^{16} or 256^2 divisions. This is common in medical images, where accuracy is important, but often 256 divisions suffice for ordinary images.

Transmit-Modulation

Transmission down any given medium is the act of sending a signal which will allow the receiver to know whether a 0 or a 1 was sent. In digital communications, this is most often done with Amplitude Modulation (AM). There is one high level of signal, and one low level of signal. Thus, we only vary the amplitude between two levels. Ideally, one might think of using an absolute 0 level, and a high level, but it is hard to shut the system off completely, so we will just refer to this as a low and high level.

The process of placing the 1's and 0's on the signal is generally called modulation. If the 1 is sent, the modulator allows the higher power signal, and if a 0 is sent there is a lower power. Thus the modulator is usually some type of on–off switch, with on for 1 and off for 0.

The tricky part is transitioning between low and high levels. This can only be done within the bandwidth constraints. The standard method for analyzing these transitions is to use the eye diagram, which we will cover later.

Medium

The medium is whatever type of physical system which you are sending signals through. In fiber optics, there are a wide variety of fibers, which we are now referring to as the medium. You are sending optical signals from a laser/modulator pair. In air, or free space, you can send nearly any type of electromagnetic signal. Radio frequencies are common but optical signals are being investigated. The general rule is that lower frequencies penetrate clouds/moisture/obstructions much better than high frequencies. Radio towers are high in order to try to gain "line of sight" communication with the user. Satellites have made "line of sight" much easier but are extremely expensive. For underwater communications with submarines, extra low-frequency (ELF) signals are used in water.

Understanding the physics of the medium you are transmitting through is fundamental to design your system, which includes the transmitter and receiver.

Receiver

The receiver is simply a microphone, in common terms, but adapted to the medium. In fiber optics, the receivers literally count the number of photons which come through the fiber. In a car, the receiver is the combination of antennae and the receiver which is internal to the radio. The receiver generally picks one channel of many, and this is all inclusive in the above diagram.

The receiver listens to find out whether a 0 or 1 was sent. The early Native Americans had digital communications long ago. A puff of smoke was a 1, and no puff was a 0. Its doubtful that these early codes were similar to modern digital communications, but a certain sequence of smoke puffs would signal some type of enemy action. Similar communication methods were also used on the Great Wall of China, and in India. With a combination of these codes, they communicated over large distances at the speed of light. While the messages may have been simple, they were effective.

Decode

The next step is decoding, or perhaps you could call it undigitizing. We have now received our 0's and 1's and we must put the sampled function values $f(k)$ back together. Thus, you could call digitizing coding and undigitizing decoding. It does not matter. Oftentimes, some more complicated version of coding or decoding is utilized if secure communications are required.

Interpolation

The final step is the interpolation, through some type of Shannon interpolation scheme. We covered ways to do this via the DFT or FFT in Chapter 5. This interpolation can oftentimes be done in hardware. Modern electronics, however, has made it possible to do this in software. If you have a digital cable box, you have probably seen the interpolation and decoding done in real time. Sometimes this happens during a thunderstorm when transmission is bad. Other times, this happens when you switch channels and the program has not caught up to the current channel.

6.1.1 Definitions from Communications

We want to introduce a number of definitions. These definitions are necessary to clearly communicate about digital communications. We must first introduce the idea of a bit time.

Definition 6.1.1 (Bit Time and Bit Period) *The time in which a single bit, which is a 0 or a 1, is transmitted is called the bit time and will be designated by β_t. The time in which an individual bit is arriving is referred to as the bit period.*

We would also like to specify another term.

Definition 6.1.2 (Bit Arrival Time) *We define the bit arrival time as the center time of a bit period.*

A simple but necessary term is

Definition 6.1.3 (Bit Rate) *The bit rate is the number of bits transmitted per second.*

Finally, we introduce

Definition 6.1.4 (Bandwidth) *The bandwidth of a communications channel is the amount of Fourier spectrum which is utilized by that channel.*

Stated simply, the bandwidth of a channel must exceed the bit rate of that channel, in standard digital communications. The numbers are generally stated differently, with bandwidth listed in Hertz, or frequencies, and bit rates in bits per second. Thus if you communicated at 5 Mbps (Megabits per second), you must use 5 Mhz (Megahertz) of bandwidth.

Similarly, the bit time is dictated by the bit rate. If one is communicating at 10 Gb (Gigabits), then the system is transferring 10^9 bits per second, and therefore, the bit time must be 10^{-9} seconds. Current fast internet service is between 5 and 25 Mb (Megabits), so the system is transmitting between 5 and 25 times 10^6 bits per second, resulting in bit times of $1/5 * 10^{-6}$ seconds or $1/25 * 10^{-6}$ seconds, respectively.

For the purposes of this book, the bit time is assumed to be constant. Advanced systems do alter the bit times according to demand, but we will assume its constant in this simple introduction. We also clarify one concept that of a symbol.

Definition 6.1.5 (Symbol) *The symbol which a digital communication system uses is just the pulse which the system transmitted. We assume that translated versions of the same pulse, or symbol, are used.*

The symbols are translated into their appropriate bit times. Thus the bit time regulates the length of the symbol.

We will also define what we call free-space communication.

Definition 6.1.6 (Free-Space Communication) *Free-space communication, or wireless communication, is the transmission through air or a vacuum (in the case of space), of a signal for communications purposes.*

Definition 6.1.7 (Synchronous Communication) *Synchronous Communication refers to a situation where both the receiver and the transmitter have a "clock," which is synchronized, so that the receiver knows exactly the time when a bit time begins and ends. Thus, the bit times are predetermined and do not need to be discovered, or tracked.*

The point is that if you are off 1/2 of a bit, the system would not generally work well. If you are off many bits, then the bit stream will be compromised and the decoding will yield false results. Many systems have safeguards to make sure that the bit stream is appropriately adjusted and the clock is kept appropriately.

Definition 6.1.8 (Asynchronous Communication) *In asynchronous communication, the clock is recovered by the receiver from the bit stream and does not need to be determined from a prearranged clock. Asynchronous communication is more adaptive but requires more from the receiver.*

We would not spend much time on protocols associated with different types of communications but rather will concentrate on the basics of communications.

6.2 Modulation

Modulation is the process of putting the actual information on the communication line. If one is sending smoke signals, the person who is fanning the flame or putting a blanket over it is a modulator. Your vocal chords, mouth, and tongue comprise the modulators for speech in your body. Modern modulators for digital communications are generally some type of switch. We will consider a couple of modulation modalities, some of which are historical and others which are relevant in modern settings.

6.2.1 Analogue vs. Digital

Traditionally, modulation was done in what is generally referred to as analogue mode. A simple walkie-talkie converts your voice to a radio frequency, sends it to corresponding radios, and then converts it back to speech. Traditional vinyl albums or records are another method of analogue recording. Small variations in the depth of the grooves on the record reflected small variations in sound, and loud variations were represented by larger changes. One problem with vinyl records is that any scratches on the surface generally would become large amplified problems.

Analogue communication is actually slightly faster than digital since it does not require the breaking down of samples into individual bits. Thus in analogue communications, the samples $f(k)$ are not broken down into their digital representations. Secondly, the signal is not really sampled but continuously recorded. This is done in the grooves of a vinyl album. Similarly, some version of the continuous voice is converted into radio waves and then reconverted back into human speech.

Digital communications involve sampling and digitization. Sampling does create a discrete number of samples which represent the continuous function. The sampling or reduction from a continuous stream of information to a discrete one is nice but does not really give the true value of digitization. The true value is when the sample $f(k)$ is digitized. The digitized sample is then sent as a 0 or a 1. If a degraded 1 signal comes through as .65, this will be reclassified as a 1. If a degraded analogue signal comes through there is little that can be done to correct the mistake, so a .65 stays .65, since there is no reason to doubt it.

The very fact that only 0's and 1's need to be detected allows one to correct the transmission with a high degree of accuracy. If we were transmitting values from 0 to 255, then we would only have 1/256'th of the ability to correct for mistakes. This would lead to many mistakes. Digital coding is a way to ensure that the entire signal is exactly transmitted. That is why digital communications are generally the preferred method for communications.

6.2.2   AM vs FM

The type of modulation which people are most familiar with exists in their car. Car radios generally operate in either AM or FM mode. In the early part of the twentieth century, AM radio was the dominant mode. In the 1960s and 1970s, radio stations were generally split between AM and FM. Currently, FM appears to be dominant because of its increased tolerance to interference. There are, however, some AM radio stations which can transmit much longer distances than the FM stations, which are generally constrained to 30–100 mile radii. Let us quickly realize what these acronyms mean.

Definition 6.2.1 (Amplitude Modulation) *Amplitude modulation (AM) is precisely what it sounds like. The signal is modulated by changing its amplitude.*

Definition 6.2.2 (Frequency Modulation) *Frequency modulation (FM) involves modulating the frequency of the signal to reflect the signal which is transmitted.*

The difference between AM and FM is shown in Figure 6.2. For the rest of this chapter, we will assume AM modulation. While FM modulation has its advantages in radio communications, AM modulation is generally used in fiber optics and most other digital communications.

6.2.3 RZ vs. NRZ

We will concentrate on AM modulation for digital communications from this point forward. There are various methods for AM modulation, however. Since we must either indicate a 0 or a 1 from the transmitter to the receiver, the typical way to do this is with an on/off switch. Off indicates a 0 and On indicates a 1. The transition between On and Off is where we place our attention and is the difference between RZ and NRZ communications.

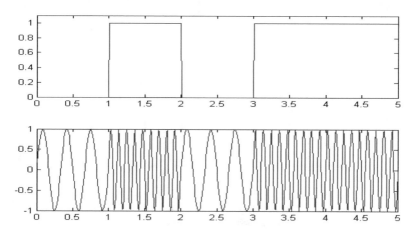

Figure 6.2: At the top, we have illustrated the AM modulation of a stream of [0,1,0,1,1]. The bottom is the FM modulated version, with the difference between a 0 and 1 denoted by different frequencies.

One way to clearly separate the difference between a 1 and a 0 is to separate the On/Off times with off times. The timing of these situations is critical, but the issue is somewhat clarified if one shuts off the transmission every time a 1 or 0 is sent. This is called return-to-zero (RZ) modulation.

Definition 6.2.3 (RZ Modulation) *In RZ communication, the system is shut off for a portion of the bit time, regardless of whether a 0 or 1 is transmitted. Thus the system "returns to zero" or RZ during every bit.*

In this situation, even if you are sending a sequence of six 1's, you would shut the transmitter off every time you send a one. If you sent a sequence [1,0,1], then you turn the transmitter on for $1/2$ of the bit time, β_t and off the for 1/2 of the bit time. The transmitter is not turned on at all during the "transmission" of the 0, and then once again it is turned on for 1/2 of the final bit, and off for 1/2 of the final bit.

This is illustrated in Figure 6.3. In NRZ, the system is not shut off between consecutive 1's. In RZ, system always "returns to zero", regardless of whether or not a 0 or 1 is transmitted.

Definition 6.2.4 (NRZ Modulation) *An NRZ modulated system does not return to zero, unless there is a 0 to be transmitted.*

Thus if you transmit six 1's, one 0, and six more 1's, the system will only shut off once, for the intermediate 0.

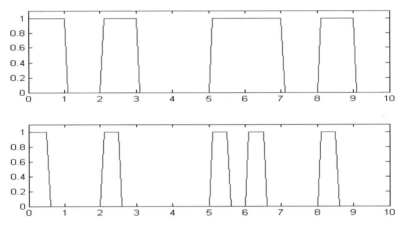

Figure 6.3: A simple illustration of RZ vs. NRZ modulation is shown above. Both signals are modulated for the bit-stream sequence [1,0,1,0,0,1,1,0,1,0]. The top is NRZ, or non-return-to-zero modulation. Notice that the signal does not return to zero between 5 and 7. The second is an RZ, or return-to-zero modulated signal. The signal goes back to zero between each bit time.

6.3 The Eye Diagram

The eye diagram is a fundamental graphical tool which is used to understand the separation of 1's and 0's in a digital communication system. Our first example of an eye diagram is given in Figure 6.4. This eye diagram illustrates the transmission of one type of pulse, $\sin^2(\pi t/2)$, with the rise and fall time shown. This pulse is shown in blue. The green lines represent shifted versions of the sinusoid, which are by nature cosines$\cos^2(\pi t/2)$. This diagram illustrates a system which is receiving pulses at the integer times 0, 1, and 2.

Notice that the top line is the sum of the sine and cosine terms. This is significant. If there was a 1 at time 0, then the greencos^2 term on the left would be turned on, and if there is a 1 at time 1, then the blue sin^2 term would be turned on. Thus if a 1 was sent at time 0 and at time 1, the receiver would receive a sin^2 + cos^2 = 1, signal from 0 to 1. This is a NRZ (non-return-to-zero) type of transmission. While transmitting sin^2 signal is sometimes not practical, it does sometimes serve as a reasonable model, which is fairly accurate for a more realistic system.

Realize that the receiver will have to "listen" for a non-trivial amount of sample time around the receive times. If the receiver were to just "sample" the output at time 0,1, or 2, then the statistical interference might cause a problem. Thus the receiver would "listen" to the signal from $1 - \epsilon$ to $1 + \epsilon$. The size of ϵ is not necessarily close to zero, but a good approximation, as shown in the picture, is that ϵ would be 1/4 of the bit time, or that the receiver would listen, and integrate the return, for 1/2 of the bit time. If this were fiber optics, then the receiver would literally count photons for 1/2 of the bit time.

The reason why the eye diagram is called what it is should be obvious. You have eyelids, as illustrated, and a center pupil/iris area. One of the most common questions in digital communications is whether the eye is open or not. If the eyelids close, it means that there is no discernible difference between a 1 transitioning to a 0, or a 0 transitioning to a 1, etc. This is illustrated in more detail in Figure 6.5.

For the bit "arriving" at time 1, the bit time is between .5 and 1.5. The blue line illustrates a single 1 which is arriving at time 1. This blue line then returns to zero at time 2. The red at top would illustrate multiple ones, as discussed above. Multiple zeros would stay at zero. The green line on the left represents a 1 at time 0, transitioning to a 0 at time 1. The green line at the right represents a transition from 1 at time 1 to 0 at time 2.

Multiple bits represented at the same time

An obvious question is "How do we tell from this diagram whether there is a 1 at time 1 or a 0?" The answer is you cannot. The eye diagram represents the aggregated transition of multiple bits, often over hours and or days. The point is that if the eye stays open, each of the individual 1's or 0's can be separated accurately. An actual eye diagram, which was formed over 17 hours at 10 Gbps is shown in Figure 6.6.

In order to form an eye diagram for a real system, we preferably take time snapshots which are two-bit lengths long. Such a time snapshot is shown in Figures 6.4 and 6.5. These snapshots are then overlayed on each other to show a moving record of how the data has been moving through the receiver. Real systems are tested not for seconds, but for minutes or hours. One of these tests is shown in Figure 6.6. Notice that Figure 6.6 is not a simple eye diagram like Figures 6.4 and 6.5. Rather it shows the statistical variation, and perhaps some of the interference between bits that occur in a system.

We emphasize the difference between RZ and NRZ transmission with the eye diagram in Figure 6.7. Notice that there are no "lower" eyelids. This is because the system returns-to-zero, for a 0. It may be somewhat confusing that the system does not exactly return-to-zero during a single bit time. There are different varieties or return-to-zero, but notice that this eye clearly demonstrated that the system does return-to-zero, for a short period of time, even if a sequence of 1's is transmitted.

6.3.1   Interference

We will now discuss some of the differences between the ideal diagrams of Figures 6.4 and 6.5 and the real eye diagram in Figure 6.6. To begin with, understand that the diagram in Figure 6.6 exhibited an error rate of 10^{-12}. At this time, we must define this concept.

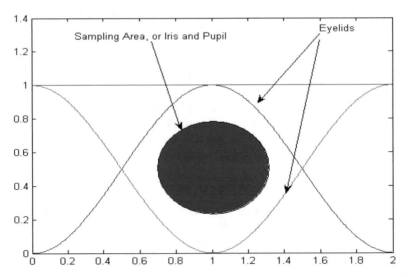

Figure 6.4:   The simple eye is illustrated above. Notice the upper and lower eyelids, the virtual pupil and iris, and the lower eyelids.

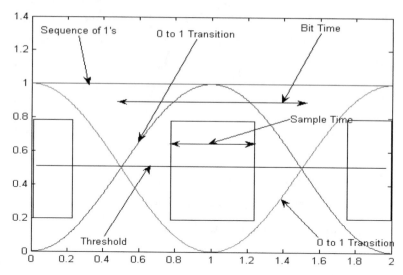

Figure 6.5: The a more technical illustration of an eye diagram is illustrated above. Notice that a sequence of 1's will remain on the top line, a sequence of 0's will remain at 0, and the transitions from 0 to 1 are clearly delineated.

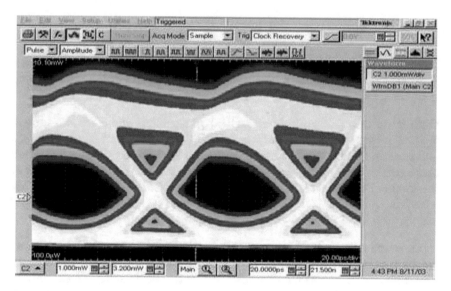

Figure 6.6: This is an eye diagram of a real 10 Gbps fiber optic channel. This was collected over 17 hours of operation. Variations in the statistics of reception, the timing and the system widen the upper and lower eyelids, but the eye remains open and the system was operating at a 10^{-12} bit error rate. The x-axis is time, and the y-axis measures power.

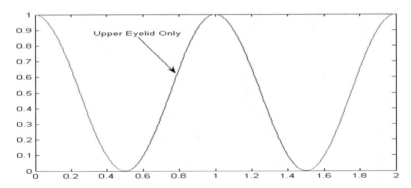

Figure 6.7: An eye diagram for RZ transmission. Note that since the system returns to zero between bits, there is no lower eye. Depending upon the type of RZ transmission, there may be a region which is zero for a while between bits. This will require more bandwidth, however.

Definition 6.3.1 (Bit Error Rate) *The Bit Error Rate of a Digital System is the number of incorrect bits, B_{wrong}, divided by the number of transmitted bits, B_{total}. Thus, we have*

$$BER = \frac{B_{wrong}}{B_{total}}.$$

The standard in most digital communications is that the BER must be less than 10^{-9} for the system to be acceptable. We will also define the following terminology.

Definition 6.3.2 (Symbols and Bits) *In communications protocols, there is often a reference to bits and symbols. Bits are simple, they are the zeros and ones that one is trying to send. A Symbol, or pulse, is the physical signal that a communication system utilizes to carry the zero or one from one end of the communication channel to the other. In prehistoric days, symbols might have been smoke puffs, or fires at night. In modern days, they are pulses of energy, usually electromagnetic, which transmit the data.*

Thus we have the symbols, and we have the digital code which is encoded on the train of symbols. Let us discuss at this time a couple of the reasons why a system may become degraded and not perform well. We define two different sources at this time.

Definition 6.3.3 (Inter-Symbol Interference) *Inter-Symbol Interference (ISI) is the interference which occurs between 0's and 1's of one channel. Namely, the effect of a 1 or a 0 on the next 1 or 0. ISI is only a problem if it creates a bit error.*

Another common term is

Definition 6.3.4 (Cross-Channel Interference) *Cross-Channel Inter-ference (CCI) is the interference between 0's and 1's of one channel, and the 0's and 1's of another channel operating at different wavelengths or fre-quencies. CCI becomes destructive when it interferes with the transmission of 1's or 0's in an adjacent channel. CCI becomes a serious problem if one channel creates a bit error on another channel.*

Simple radio stations and cell phones often give recognizable examples of ISI and CCI. When tuning a radio, oftentimes if the signal is weak you can hear two different radio stations at one time. This is generally CCI, with a channel at a different frequency "leaking" into the channel you want to hear. This is fairly common in AM radio.

There are also times when on a cell phone, you begin to hear the conversa-tions of persons on another channel, or perhaps on your frequency, but with ISI interference because the bits you are using are similar. While this could be CCI, this is probably to be ISI, where the "clock" of the receiver phone is off, and therefore hears two phones at the same time. There are many other effects which cause problems with digital communications systems. We will concentrate on these in this book since they are common and must be addressed before any system can operate.

6.4   Fourier Constraints

The interaction of Inter-Symbol Interference and Cross-Channel Interference is very enlightening when trying to understand the constraints placed on a communication system. The most important thing to remember about a communication system is that bandwidth is very expensive. Thus any system is a compromise between performance and cost.

In free-space communications, or when communicating with a wireless device, the Federal Communications Commission (FCC) regulates the amount of bandwidth that a user, whether a radio station or wireless phone provider, can utilize. The user of wireless airspace in any area pays for the right to a certain frequency region. The revenues from these leases amount to billions of dollars which are paid to the FCC. Utilizing the airspace appropriately and efficiently is central to making a communications company viable. The FCC requires that the channels you are operating do not interfere substantially with another users' channel, for obvious reasons.

Cable, which generally consists of multiple copper wires, is expensive to install. While it is currently deployed to nearly all households in the USA, the cost of stringing and maintaining that cable is enormous. Cable companies are continually trying to figure out ways to provide more channels, more pay-per-view movies, and higher speed internet to the end user.

Fiber optic cables allow an incredible amount of bandwidth to the end user. The expensive equipment required to provide this bandwidth has slowed the widespread use of fiber-to-the home in most places. We may see this change, but the issue is still the same, utilizing the available channels with minimal cost and maximum bit rates.

Bandwidth must be paid for, either through leases to the FCC or another third party, or by laying cable or fiber optics for your own system. Thus bandwidth is expensive.

6.4.1   Uncertainty, Dilation, and ISI

Let us recall the NRZ eye diagrams in Figures 6.4, 6.5, and 6.6. For the first two of the diagrams, we utilized a simple $\sin^2$ function, which was zero at the prior bit time, 1 at the center of bit region of interest, and 0 again at the following bit time. If we contrast this with the RZ eye diagram, we see that the $\sin^2$ function for the RZ system has half the length of that for the NRZ system.

Remember the relationship between dilation and the Fourier transform, namely from Theorem 4.7.9 we have

$$\mathcal{F}(f(at)) = \frac{1}{a}\hat{f}(\frac{s}{a}).$$

Thus, when we make a pulse half as long, then its Fourier Transform must be twice the size of the original pulse. Thus the RZ system requires twice as much bandwidth as an NRZ system.

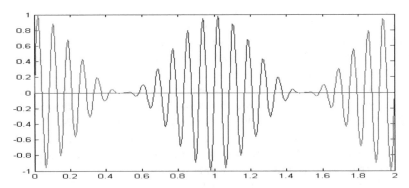

Figure 6.8:   We illustrate more realistic pulses, which might be used for a communications system. Oftentimes the frequency is so fast, relative to the bit length, that only the amplitude of the peaks is illustrated in the eye diagram.

Returning once again to Figures 6.4 and 6.5, realize that we arbitrarily made the bit arrival times at the integers, 0, 1, and 2. If we want this channel to operate at twice the bit rate, then the arrival times will have to be at 0, .5, 1, etc. As a result, we will have to scale the function which we are using to be 1/2 the original length. This will require twice the bandwidth.

Now, to minimize Inter-Symbol Interference (ISI), we need to make sure the pulses do not run into each other. One way to do this is to shorten the length of the symbol, or transmitted pulse, as we did for RZ in Figure 6.7. This will directly increase the bandwidth of the channel. Remember that bandwidth costs money, so this simple solution may lower the ISI, but it comes at a definite cost. In general we have a rule,

Definition 6.4.1 (Communications Rule 1) *Shortening the pulse or symbol will minimize the ISI and increase the bandwidth.*

6.4.2 Uncertainty, Dilation, and CCI

Now we must consider the Cross-Channel Interference (CCI). To do this, we must first rethink the reality of our eye diagrams. The pulses which generate the eye diagrams of Figures 6.4, 6.5, and 6.6 are very simple. Generally, one transmits a pulse whose frequency is very fast relative to the length of the pulse. This is illustrated in Figure 6.8.

It is often convenient to use a representation of a window, which is the amplitude, times a pure frequency or sine or cosine to designate these type of pulses. Thus our pulses of choice are

$$p(t) = w(t - t_b)\sin(2\pi f_k t),$$

where t_b is the center bit time, and f_k will end up being the center frequency.

Each channel will correspond to a different center frequency f_k. Two such pulses, and their center frequencies are shown in Figure 6.9. Notice that the third graph in Figure 6.9 is symmetric. This is always the case when the absolute value of the Fourier Transform of a real-valued function is displayed. As a result, oftentimes only one side of the graph is displayed.

In this type of representation, one channel would consist of shifted versions of the window times one frequency, or sinusoid, (the symbol for that channel), and the next channel would consist of shifted versions of the window times a second frequency function, or sinusoid.

We mentioned before that unless you are utilizing a copper cable or fiber optic which is exclusively controlled by your company, there are severe restrictions on the bandwidth you can use and the amount of frequency content that can overlap into someone else's channels. Either the FCC or by the owner of the fiber or cable will enforce these regulations. To increase the speed of your channel, you must decrease the length of your pulses, and therefore increase the amount of bandwidth you are using. This is the compromise which restricts the speed of any communications channel. An example of this type of channel is shown in Figure 6.10.

We must return to dilation and uncertainty to clearly understand the above discussion. If we decrease the length of our pulse, we increase the bandwidth, and as a result increase the CCI. On the opposite side, we have a second communications rule.

Definition 6.4.2 (Communications Rule 2:) *To minimize CCI, one should increase the length of the transmitted pulse or symbol, which decreases the associated bandwidth.*

Thus we have a dilemma. If we are to minimize ISI, Communications Rule 1 tells us to shorten the pulse. If we are to minimize CCI, Communications Rule 2 tells us to lengthen the pulse. These are the two constraints which must be met in order to have multiple channels operating in one medium. The pulse must be short enough to allow ISI to be insignificant, and long enough to make sure that CCI is not significant. This limits the bit times and the bandwidths of the system.

6.5 Statistics and Matched Filtering

Previously, we have examined deterministic ways and constraints involved in the design of a digital communications system. We will now examine the statistical problems involved in such a system, and the uncertainty which these random variations cause.

When we talk about the deterministic problem, we are assuming that we can transmit exactly the pulse which we choose, and we know exactly how that pulse will arrive at the other end of the system. Even if you have an exact knowledge of the medium, say in fiber optics, there are variations in the current levels of your modulator, the temperature of the repeater stations, and small imperfections caused when the fiber was laid in the ground.

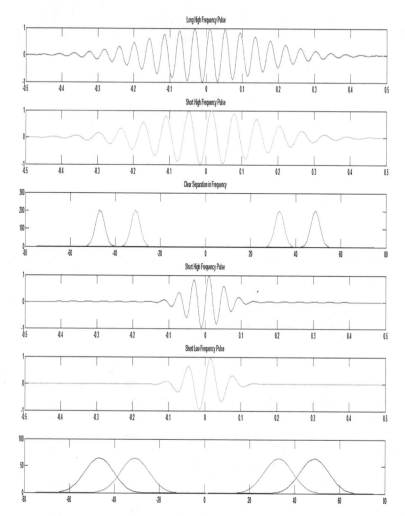

Figure 6.9:   We illustrate four pulses and their spectra above. The first and second graphs illustrate two pulses of the same length and different frequencies. The third graph is the Fourier transform of the pulses, with the higher frequency pulse illustrated in blue, and the lower frequency pulse illustrated in green. This shows that they are separated in frequency so a frequency filter can clearly pull them apart. The fourth and fifth graphs illustrate the same frequency pulses, but with shorter time lengths. The sixth graph shows that dilation, or the uncertainty principle, has made their spectra overlap so that they cannot be separated merely by frequency.

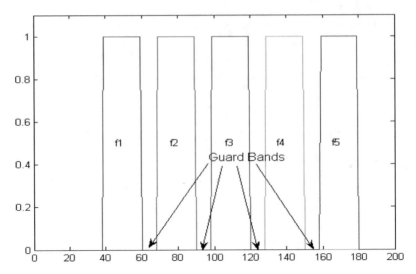

Figure 6.10: We illustrate the standard setup for a communications channel above, with
five frequency windows, and guard channels between them. Generally, the FCC or the owner
of the cable/fiber place restrictions on how much power can "leak" from one channel to
the next. As a result, guard channels are used to insure that there is no Cross-Channel
Interference (CCI).

In addition, there are statistical variations due to the nature of the photons which are transmitted by the laser. In free-space communications, the
humidity, temperature, cloud cover, etc., will affect the signal which is sent.

    We could try to describe all of these problems in a very detailed way.
Much effort is made to try to isolate each one of these problems. Oftentimes,
however, it is easier and helpful to put all of these problems into one basket
and call them random statistical variations.

    The most typical way to do this is by using a statistical model, such as

$$r(t) = s(t) + n(t),$$

where $r(t)$ is the received signal, $s(t)$ is the transmitted signal, and $n(t)$ is
the statistical noise associated with all of the above random variations in
conditions. This model is rarely exact, but is useful in trying to develop a
simple way to deal with the very complex number of interference terms which
one will encounter.

    The question now is "what do we mean by n(t)?" That is also part of the
statistical model. Namely, we assume that the noise is generated by a process,
which hopefully is consistent with the reality that we observe.

6.5.1 Gaussian White Noise

A standard model for noise is Gaussian White Noise. We will now quickly define these terms. In general, those who are not at all familiar with statistics should refer to any standard statistics book. Detailed knowledge of statistics is not required, however. We will try to outline the specific knowledge which is required here.

Definition 6.5.1 (Gaussian Distribution) *The Gaussian distribution function is defined by*

$$f(t) = \frac{1}{\sqrt{2\pi\sigma^2}} e^{-(t-\mu)^2/(2\sigma^2)}.$$

This distribution function determines a random Gaussian process $n(t)$, which can be sampled at any time t. The process is tied to the distribution by the probability property that the probability that any sample $n(t)$ lies in an interval $[a, b]$ is given by

$$P[a < n(t) < b] = \int_a^b f(t)dt.$$

This distribution and process is defined to be $N(\mu, \sigma)$. We will refer to μ as the mean and σ as the standard deviation. We will also use the notation $n(t) \sim N(\mu, \sigma)$ to denote that the distribution associated with $N(\mu, \sigma)$ are the above normal distribution.

We will begin by illustrating the distribution of $N(2, 1)$. Notice that by the definition of the process, the values near 2 are "more likely" than the values near 4. In addition, we would like to consider the most likely value. This is defined by the mean, or expectation, which is

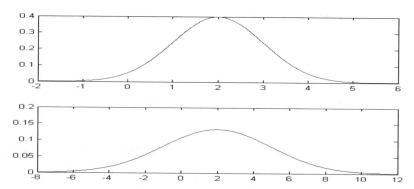

Figure 6.11: We illustrate the distribution function $f(t)$ for $N(2, 1)$, and $N(2, 3)$ above. Notice that by the definition of the process, the probability of having results near 2 is higher than anywhere else. The second, or wider distribution $N(2, 3)$, illustrates that with a larger variance, the likelihood of having values far from 2 is increased. With the first distribution, it seems unlikely that you would get a result outside of $[-1, 5]$. The second distribution seems to allow values in the region $[-8, 12]$.

Definition 6.5.2 (Mean) *The mean or expectation of a continuous distribution $f(t)$ is given by*

$$E(f(t)) = \int tf(t)dt.$$

It can easily be shown, see exercise 1, that $E(N(\mu, \sigma)) = \mu$. The other property illustrated in Figure 6.11 is the variance. The variance measures how wide the distribution is, or how far from 2 you can reasonably expect to get results from the process. A small variance implies that there is not much noise, and nearly all returns will be "near" 2.

We now want to address the idea of the variance, or the second term σ^2 in $N(\mu, \sigma)$.

Definition 6.5.3 (Variance) *The variance of a process with a probability distribution $f(t)$ is given by*

$$\sigma^2(f) = \int (t - \mu)^2 f(t)dt.$$

The standard deviation is defined to be σ.

In general, the noise associated with a communications system is assumed to have mean 0 and variance σ^2, or be $N(0, \sigma^2)$. In addition, we assume that the noise sampled at any two separate times is unrelated, or independent, except that it comes from the same distribution. The general term for this is Independent Identically Distributed (IID), but that is probably more than necessary. Another general terms are that this is Gaussian, white noise, where white stands for uncorrelated.

6.5.2   Averaging Noise

If different samples from a IID process are taken, then their sum $S = \sum_{k=1}^{N} n_k$ is also a random variable. If these are normal random variables, then the sum is also, and we have $E(S) = \sum_{k=1}^{N} E(n_k)$, since the expectation is linear. If they all have mean μ, then we obviously have $E(S) = N\mu$. Moreover, if they are independent, then the variance also adds, so $\sigma^2(S) = \sum_{k=1}^{N} \sigma^2(n_k)$. If they are samples from $N(0, \sigma)$, then we have $\sigma^2(S) = N\sigma^2$.

Now the question becomes "How does the variance, or standard deviation of a variable change under multiplication?". In other words, how is the variance of $cn(t)$, or $\sigma^2(cn(t))$ related to $\sigma^2(n) = \sigma^2$. The answer is that $\sigma^2(cX) = c^2\sigma^2(X)$. While this makes some sense notationally, the proof is enlightening and is left as an exercise. To see it visually, consider Figure 6.12. We have plotted two normal variables, $N(0, 1)$, and $N(0, 2)$. If $X \sim N(0, 1)$, then $2X \sim N(0, 2)$. Thus the standard deviation is changed by the multiple, and the variance by its square. A continuance of this argument yields the following.

Theorem 6.5.1 *If X_k are independent random variables, then*

$$\sigma^2\left(\sum_k c_k X_k\right) = \sum_k c_k^2 \sigma^2(X_k).$$

First of all, lets think about the variance of an average. Thus, if we average noise, i.e., we consider $(\sum_{k=1}^{N} n(t))/N$, then the variance is

$$\sigma^2\left(\sum_{k=1}^{N} n(t_k))/N\right) = \frac{1}{N^2}\sum_{k=1}^{N} \sigma^2(n(t_k)) = \frac{1}{N^2}N\sigma^2 = \frac{1}{N}\sigma^2,$$

where we assume that the $n(t) \sim N(\mu, \sigma^2)$ for all t. Thus averaging noise yields noise with the variance cut by a factor of $1/N$, or the standard deviation cut by a factor of $1/\sqrt{N}$. More generally, we are interested in inner products of a vector of constraints $\vec{c}$, such that $\|c\| = 1$, and a sequence of random variables. By the same argument as above

$$\sigma^2(\vec{c_k} \cdot n(\vec{t_k})) = \sum_k c_k^2 \sigma^2(n(t_k)) = \left(\sum_k c_k^2\right)\sigma^2 = \sigma^2.$$

Thus the inner product of a vector $\vec{c}$ with length one against a sequence of random variables which are $N(\mu, \sigma)$ does not change the variance. More generally, we have shown

$$\sigma^2(\vec{c_k} \cdot n(\vec{t_k})) = \|c\|^2 \sigma^2.$$

Note that this includes both of the above examples.

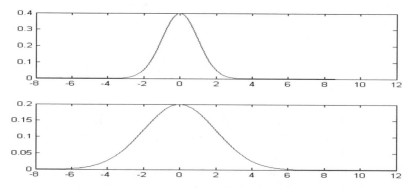

Figure 6.12: We illustrate two distribution functions above. At the top, we have $N(0, 1)$, and at the bottom, we have $N(0, 2)$. Note that if $X \sim N(0, 1)$, then $2X \sim N(0, 2)$.

Signal-to-Noise Ratio

One of the most important measures of the performance of a system is the Signal-to-Noise Ratio. Building a communications system is primarily an issue of managing and manipulating the signal-to-noise ratio. There are different possible definitions, but since we are using the L^2 metric almost exclusively in this book, we define it as

Definition 6.5.4 (Signal-to-Noise Ratio (SNR)) *The Signal-to-Noise Ratio is given simply by*

$$SNR = \frac{\|Signal\|_2}{\|Noise\|_2}.$$ (6.1)

Oftentimes, the logarithm of the SNR or

$$\log_{10}(SNR) = \log_{10}\left(\frac{\|Signal\|_2}{\|Noise\|_2}\right)$$ (6.2)

is used. There should be no confusion, because when the logarithmic SNR is used, it is stated in dB, and absolute terms are used for the simple SNR

The manipulation of the SNR and the resulting differences in the SNR is very important in understanding and designing communications systems.

6.5.3 The Matched Filter

Now let us return to our model, where we have a received signal $r(t) = s(t) + n(t)$. We will assume that we have a finite number of samples of this, say $r(t_k)$. Our goal is to maximize the amount of information we get from $s(t)$, while minimizing the effect of $n(t)$. Remember that $s(t)$ tells us whether we have a 1 or a 0, and $n(t)$ is just noise which is in our way.

We will consider an arbitrary filter f_k, which is the same length as $r(t_k)$. Thus, we are analyzing the inner product

$$\vec{f} \cdot r(\vec{t_k}) = \sum_k f_k r(t_k)) = \sum f_k s(t_k) + f_k n(t_k) = \vec{f} \cdot s(\vec{t_k}) + \vec{f} \cdot n(\vec{t_k}).$$

Assuming that we have $n(t) \sim N(0, \sigma)$, it follows that the term $\vec{f} \cdot n(\vec{t_k}) \sim N(0, \sigma)$. Therefore, the variance of the sequence will not change regardless of the choice of $\vec{f}$.

The mean of this inner product, however, will be given by the non-random portion, $\vec{f} \cdot s(\vec{t_k})$. Assuming that $s(t_k)$ represents a 1 at some point in time, we want this mean to be as large as possible so that we can tell a difference between this 1 and other 0's. Therefore, we want to maximize $\vec{f} \cdot s(\vec{t_k})$. Since changing f will not alter the variance, this is the appropriate thing to do, maximize the first portion, and the second portion stays constant.

This is very easy, however. Recall that the Cauchy–Schwartz inequality tells us that $|\langle f, g \rangle| \leq \|f\|\|g\|$ with equality happening only when $f(t) = cg(t)$. The unique filter f which maximizes the above is

$$f = \frac{s(\vec{t_k})}{\|s(\vec{t_k})\|},$$

where we divide by the length of $s(\vec{t_k})$ to make f have length one.

While this seems like a final solution, this would mean that we need a separate filter for each of the individual bits. That is a possibility, but instead, we usually do this with convolution. Recall that convolution is simply a series of shifted inner products. Our symbols or the pulses which carry the 0's and 1's are also just shifted copies of one another. Therefore, we can just take a copy of our pulse, and the convolution will yield the maximum values at the appropriate (shifted) times.

Theorem 6.5.2 (Matched Filter) *Let $s(t)$ be a function which consists of a localized and shifted function $f(t - t_k)$, for some t_k with $\|f\| = 1$. Let $r(t) = s(t) + n(t)$ with $n(t) \sim N(0, \sigma)$. Then the collection of random variables*

$$r * f(t) = \int r(\tau) f(t - \tau) d\tau$$

will be a sequence of normal random variables with a maximum mean of 1 at t_k, and variance σ^2 for all t. In addition, we refer to $f(t)$ as the matched filter for the system. There is no other function other than $f(t)$ which will meet this performance criterion.

The simple example in Figure 6.15 shows that matched filtering is not only a useful technique, but it is also absolutely necessary to optimize the performance of a communications system. The goal is to try to reduce the effect of the noise, while boosting the effect of the signal. There are many theorems other than the one presented here, which show that statistically this is the right thing to do.

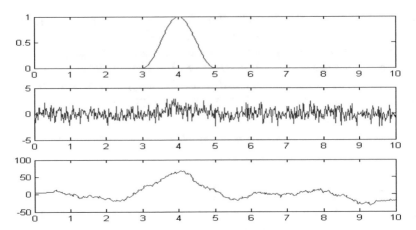

Figure 6.13: At the top, we illustrate an NRZ bit sequence with a 1 at location 4, and zeros at the other integers. The middle is an illustration of relatively high noise, taken from a $N(0, 1)$ distribution. It is difficult, from visual inspection to tell where the 1 is and where the 0's are. The matched filtered version is shown at the bottom. It is far easier to locate the zeros and one in this image.

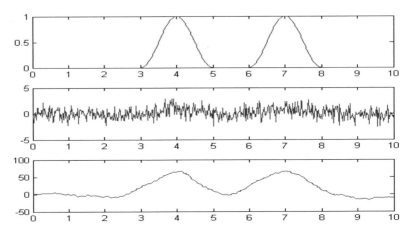

Figure 6.14: At the top, we illustrate an NRZ bit sequence with a 1 at location 4, and 7 and zeros at the other integers. The middle is an illustration of a lower noise level, a $N(0, 1/2)$ distribution. The ones are still difficult to see in the original. The bottom is the matched filtered version. It is easy to tell by visual inspection where the 1's are and where the 0's are.

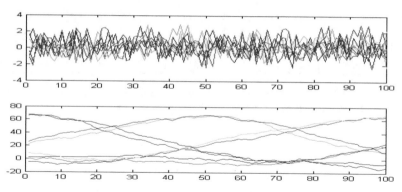

Figure 6.15: The eye diagram from the unfiltered result of Figure 6.14 is at the top. It is not usable. The eye is not even visible, and clearly closed. The eye diagram constructed from the filtered result of Figure 6.14 is shown at the bottom. Note the two ones, at the top of the eye diagram. The upper eyelids are from the two transitions from 1 to 0, and the lower eyelids are similar transitions. While this eye is not pretty, it is open and would work.

6.5.4   Transmitter and Receiver Details

The diagram we introduced at the beginning of the Chapter in Figure 6.1 is very simplistic. We would like to expand at this time upon the transmitter and the receiver.

Transmitter

To begin with, we introduce Figure 6.16 which is an expanded view of the transmitter. The pulse generator generally creates a sequence of pulses, which are all 1's at this point in time. These pulses are then sent to the modulator. The modulator is an on–off switch which when on, let's the 1's pass as 1's. When a zero needs to be inserted, the switch turns that particular pulse off creating a zero in the pulse stream.

Finally, there is the filter. This is usually not a matched filter, but rather a narrowband filter which ensures that the bandwidth of the channel is not exceeded. Referring to Figure 6.10, this filter makes sure unwanted frequencies which may have been introduced along the way are eliminated. Thus the pulse stream stays within the channel specifications. The filter can also be used to shape the pulses. This is particularly true in fiber optics, where electronic measures to shape the pulse are not sufficiently fast.

Receiver

Similarly, we illustrate an expanded diagram of a receiver in Figure 6.17. There is often some type of amplifier to increase a weak signal at the receiving end of the material.

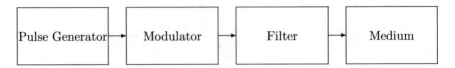

Figure 6.16:   A basic transmitter diagram is shown above. The first three components, the pulse generator, modulator, and filter, are all a part of the transmitter.

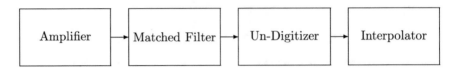

Figure 6.17:   A basic receiver diagram is shown above. All of the components are necessary to optimize the receiver.

The second step is generally to use a matched filter to manage the noise. If this works properly, a correct eye diagram will be produced, and the result will be 0's and 1's, which are correct. These have to be converted back into samples $f(k)$, and then interpolated into the signals, or images which we were trying to transmit.

6.5.5   Chapter Project:

1. **Simulating an RZ system** Construct a sequence of RZ pulses which are 64 pixels long, resulting in a signal which is 2048 pixels long in MATLAB. You can use either a Gaussian which is essentially 64 pulses long, or a raised cosine, or another function. Then create an eye diagram from this sequence. This is done by placing all of the pulses in an array which is 64x32.

2. Use a sequence such as above to create an eye diagram, and add noise to the signals at different levels, to see when the eye diagram closes. When the eye closes, the system will no longer work.

3. Have MATLAB create a pseudorandom set of 0's and 1's. Use these to turn simulate a modulator turning the pulses in Problem 1 on and off, by multiplying the off pulses by 0, and the on pulses by 1. Once again add noise at different levels and check out when the eye diagram is open and when it is closed.

4. Use filtering, such as nonlinear or linear filtering from Chapter 3 to improve this result.

5. **Simulating and NRZ system** Construct a sequence of NRZ pulses which are 64 pixels long, resulting in a signal which is 2048 pixels long in MATLAB. You can use either a Gaussian which is essentially 64 pulses long, or a raised cosine, or another function. Then create an eye diagram from this sequence. This is done by placing all of the pulses in an array which is 64 × 32.

6. Use a sequence such as above to create an eye diagram, and add noise to the signals at different levels, to see when the eye diagram closes. When the eye closes, the system will no longer work.

7. Have MATLAB create a pseudorandom set of 0's and 1's. Use these to turn simulate a modulator turning the pulses in Problem 5 on and off, by multiplying the off pulses by 0, and the on pulses by 1. Once again add noise at different levels and check out when the eye diagram is open and when it is closed.

8. Use filtering, such as nonlinear or linear filtering from Chapter 3 to improve this result. You can also try matched filtering.

Chapter 7
Radar Processing

Radar processing began in most countries in the 1930s. The USA had many operational radar sets at the time of Pearl Harbor. The radar on Oahu did detect the Japanese attack, but the communications with the central command was very slow. Realize that the navy had sunk a submarine before the air attack and even that did not awaken the commanders in Hawaii.

Similarly, Britain, Russia, and Germany had operational radars of some type during the war. Only Japan seriously neglected this as a method to optimize the use of limited air power to defend themselves.

7.1 Range and Velocity Measurements

7.1.1 Range Measurements

The two most basic measurements in radar are range and doppler. Range is simply the measurement of the distance of the target from a single (monostatic), radar antennae. Generally the antennae emits a radar signal, or pulse, of some type, and then listens for the "echo." This is depicted in Figure 7.1.

Putting things into mathematical form, let us define the terms R, T_R, and c.

Definition 7.1.1 (Radar variables) *Referring to the diagram in Figure 7.1 we define the distance from the antennae the target (the airplane in the diagram) to be R. The time it takes a transmitted pulse to "echo" off the target and return to the antennae is T_R, and the speed of the pulse (generally the speed of sound) in the medium is c.*

© Springer Science+Business Media, LLC 2017
T. Olson, *Applied Fourier Analysis*,
https://doi.org/10.1007/978-1-4939-7393-4_7

The diagram in Figure 7.1 is equally valid in sonar, but the pulses would generally be sound and the speed c would be dramatically different, i.e., the speed of sound in water, not the speed of light in air.

Our first equation from Figure 7.1 is simply

$$T_R = \frac{2R}{c}. \tag{7.1}$$

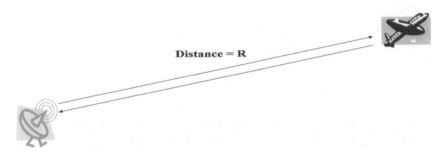

Distance = R

Figure 7.1: Above is a schematic diagram for a typical radar system. A pulse is sent, must travel a distance R to the target, and then return back to the antennae. The round trip distance is therefore 2R.

In English, the distance traveled by the pulse divided by the speed of travel is the time traveled. This gives us our range estimate

$$R = \frac{cT_R}{2}. \tag{7.2}$$

The impression from Figure 7.1 and equations 7.1 and 7.2 is that radar is easy. One certainly doesn't need Fourier Analysis or advanced mathematics. This might be true in a very ideal environment, but the details become very complicated when multiple objects are in the field of view, and one wants to distinguish amongst them.

7.1.2 Simple Range Resolution

In radar, there is one fundamental problem. The speed of light is 3×10^8 meters per second. Thus, a mistake of .01 seconds in timing the return pulse, or in recording T_R means that we will be off by 3×10^6 meters, or 30 kilometers in our range estimation. This gives us some understanding of why radar was not used until the 1930s and after. We would like to have resolution on the order of meters, for estimating an attack on Pearl Harbor. This requires timing resolution on the order of 10^{-8}, or 10 picoseconds.

Let us create a simple mathematical formula which defines the range resolution.

Definition 7.1.2 *If the uncertainty in the return time T_R is given by ΔT_R and the uncertainty in range is given by ΔR, then we have the simple relationships*

$$\Delta R = \frac{c\Delta T_R}{2} \qquad (7.3)$$

and

$$\Delta T_R = \frac{2\Delta R}{c}. \qquad (7.4)$$

Thus, if we decide on a range resolution we would like, say one kilometer (which is very large by modern standards), then we must divide by $c = 3 \times 10^8$, and we end up having to have a timing resolution of $.66 \times 10^{-5}$ seconds. For a more useful system, we want a resolution on the order of a meter, or 1.5 meters for arithmetic sake, which means we need a time resolution of 10^{-8} seconds, or 10 nanoseconds.

Thus, any ideas of using a typical stopwatch to judge the return time have to be given up. The return time must be recorded in an automated way, with the help of advanced electronics and an advanced electronic clock.

7.1.3   Maximum Unambiguous Range

Generally in a radar system, we need to send multiple pulses. Thus, there is a time between pulses, which we define to be T_p. We do not want pulses from targets at different ranges returning at the same time. Thus, we must let our first pulse clear the area which we want to scan, before sending another pulse. This gives us the idea of maximum unambiguous range, derived from 7.2 and 7.3 as

$$R_{un} = \frac{cT_p}{2}. \qquad (7.5)$$

This also lets us derive what we will refer to as the pulse repetition frequency (PRF), given by

$$PRF = \frac{1}{T_p}.$$

One problem with this idea is that you cannot control where your targets are. Therefore, if there are targets beyond the maximum unambiguous range, their returns will potentially still return at the same time as a much closer target. We will address this with the idea of the Radar Range Equation, which we explore in the next section.

7.1.4 Velocity Estimation and Resolution

The estimation of velocity is very simple, once one believes that they have a good estimation of range. Thus, we have that

$$V(t) = \frac{d}{dt}R = \lim_{\delta \to 0} \frac{R(t+\delta t) - R(t)}{\delta t} \approx \frac{R(t+\delta t_0) - R(t)}{\delta t_0}, \qquad (7.6)$$

as long as δt_0 is appropriately small. Now the question is "How reliable is the velocity?," or how reliable is this. Allowing our former notation,

$$\Delta V \equiv \frac{\Delta R_{t-\delta t_0} - \Delta R_t}{\delta t_0} \approx \frac{d^2}{dt^2}R(t). \qquad (7.7)$$

If δt_0 is large, this is a good estimation of the average velocity. To make it a good estimation of the actual velocity, you have to have $\delta t_0 \to 0$, which makes this a very unstable estimate.

Why is this an unstable estimate at $\delta t_0 \to 0$? This depends upon the resolution of the range, and the size of δt_0. If the range resolution ΔR is very large and δt_0 is small, then ΔV is expected to be large, and the accuracy of any range calculation done in this matter is not good. We will explore better ways to do this as we go on.

7.2 Radar Range Equation

We will now address basic issues of detectability. This issue is generally referred to as the Radar Range Equation, which gives us an answer to the question "How far away can I expect my radar to detect an object, or target?". Understand that the common phrase for a object in military terms is a target. Whether this is a friendly aircraft, or truly a target, which might be an enemy tank or plane, the term target is used generically.

Since by definition, a radar system is a remote detection device, this question of how remote is fundamental. There are a couple of questions which are interlocked in this process.

1. How much power will eventually strike the target?

2. How much power will eventually return from the target to the antennae?

3. How much returned power to the antennae will be enough?

The first two problems are very similar. The third depends heavily upon the nature of the radar system. The fundamental idea dominating the first two questions is really, "How does the power emanating from my antenna, or from the target, get disbursed or spread?". Let us first define P_t to be the amount of power transmitted from the antennae. In the simplest of worlds, an antenna only radiates a pulse from a point source which is not focused. The power from that pulse is then evenly spread over a sphere of increasing radius. The area of a sphere is $4\pi R^2$, so the most basic equation is

$$P = \frac{P_t}{4\pi R^2},\tag{7.8}$$

where P is the power per unit of area on a sphere of radius R, or at a distance R from the antenna.

In reality, antennae are focused to some degree or another. Note that the amount of focus is generally dependent upon the size of the antenna. A large antenna can be very well focused, but a small antenna cannot be focused to the same degree. One way or another, an antenna will illuminate some portion of a sphere at distance R from the antenna. This focusing is called the antenna gain, or G, and therefore, our equation moves to

$$P_{target} = \frac{P_t G}{4\pi R^2},\tag{7.9}$$

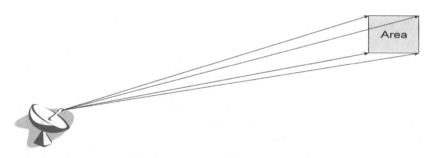

Figure 7.2:   We illustrate the focus factor and the gain of the antenna above. If the antenna only irradiates a small portion of the sphere, as above, then it has a large gain factor.

where G is the increase in power on that portion of the sphere where the antenna is focused. Thus if the antenna illuminates only 1/20'th of the sphere, G would be 20. This is illustrated in Figure 7.2. The formula for the gain is $G = (4\pi R^2)/A$, where the area or A is as illustrated. Thus, a well-focused antenna has a small illuminated area and a large gain G.

Now, the second question of how much power returns from the target. The answer is almost the same, except that we do not know the properties of the target. Secondly, we don't know, ahead of time, the orientation of the target. A plane traveling sideways will reflect the radar signal very differently than one traveling directly toward the antenna. Thus, we use an arbitrary factor σ to reflect what we call the radar cross section (rcs) of the target. This is equivalent to the radar gain factor for the antenna. Stealth fighters and bombers try to make σ as small as possible, by absorbing the radar signal, or reflecting it in as many different directions as possible. Most importantly, the attempt is made to not radiate the signal back to the antenna. Our second equation is therefore

$$P_{antenna} = \frac{P_{target}\sigma}{4\pi R^2}. \tag{7.10}$$

The equations 7.9 and 7.11 can be combined, to get the final form

$$P_{antenna} = \frac{P_t G \sigma}{(4\pi R^2)^2} = \frac{P_t G \sigma}{16\pi^2 R^4}. \tag{7.11}$$

The final question is "How much power at the antenna is enough for detection?". We will dodge this for now, and just define P_{min} to be the minimum power which is necessary for an antenna to detect an object. Thus, we define the maximum distance R_{max} in terms of the minimum power by

$$P_{min} = \frac{P_t G \sigma}{(4\pi R^2)^2} = \frac{P_t G \sigma}{16\pi^2 R_{max}^4}, \tag{7.12}$$

which we solve for R_{max} to get

$$R_{max} = \left(\frac{P_t G \sigma}{16\pi^2 P_{min}}\right)^{1/4}. \tag{7.13}$$

We need to note several things at this time. First of all, we will consider equation 7.11. If we double the distance to the target, then we don't cut the power by a factor of 2. Inserting $2R$ for R, we see that the power is cut by a factor of 16. Thus, increasing distance involves the factor R^4 in the denominator. Increasing the distance by a factor of 4 results in a power loss on the order of 256. Thus, power is of the utmost importance. This is a limiting factor in nearly all radar systems. To have an effective long distance radar, you need lots of power. The constraint is often that this power will end up frying the very sensitive components which are necessary to amplify the return signal. Isolating the sensitive components form the high power transmission signals is a large part of the art of radar design.

7.2.1   Problems and Exercises:

1. (a) What should the pulse repetition frequency of a radar in order to achieve a maximum unambiguous range of 60 kilometers?

(b) How long does it take for the radar signal to travel out and back when the target is at the maximum unambiguous range?

(c) If the radar has a pulse width of 1.5×10^{-8} seconds, how far apart in meters must equal-sized targets be separated in order to make sure that they have separate radar returns, or that they can be independently identified?

2. The moon is a radar target that may be described as follows: average distance to the moon is 3.88×10^8 meters; its experimentally measured radar cross section is 6.64×10^{11} meters squared, and its radius is 1.739×10^6 meters.

(a) What is the round-trip time in seconds of a radar pulse to the moon and back?

(b) What should the pulse repetition frequency (PRF) be in order to have no range ambiguities (Related directly to the maximum unambiguous range)?

(c) For the purpose of probing the nature of the moon's surface, a much higher PRF could be used than that found in (b). How high could the PRF be if the purpose is to observe the echoes from the moon's front half?

3. A radar mounted on an automobile is meant to determine the distance to a vehicle traveling directly in front of it. The pulse width of the radar is 10^8 seconds, and the maximum range is to be 500 meters.

(a) What is the pulse repetition rate corresponding to this distance?

(b) What is the range resolution in meters?

(c) Find the power required to detect at 5 meter square radar cross section vehicle at 500 meters, if the minimum detectable signal is $5 \times 10^{-13)}$ Watts.

7.3 Signal to Noise Management in Radar

Radar engineering is focused primarily upon being able to detect the smallest target possible, at the greatest distance, with the currently available power, antenna, and processing. Once the hardware is determined, there are a number of steps which are utilized to improve the Signal to Noise Ratio (SNR), and in essence, increase the distance by decreasing the necessary P_{min} from above. Thus the goal, given a transmitter, receiver, and antenna, is to be able to detect signals which are very small.

7.3.1 Matched Filter

One of the first tools used is the matched filter, which we introduced for communications in Section 6.5.3. When the signal is received, a radar antenna is not all that much different than the receiver in a communications. The goal is to maximize the distance transmitted. Remember in Section 6.5.3, we utilized the matched filter to maximize the signal contribution while minimizing the effect of noise. This is the same in radar processing, for the exact same reasons. The problem is having enough power to see the target, while the noise in the system will become large.

There are many sources of noise in radar. If one is attempting to image airplanes, clouds can become a problem. Generally one selects different frequencies for different applications. A typical weather radar uses frequencies which do bounce off of clouds for obvious reasons. If the goal is imaging planes, it is desirable to have the frequencies penetrate water vapor, and only bounce of hard objects. This is not that easy. While there are better and worse frequencies for different applications, it is not generally possible to have one frequency target clouds, and another planes with not overlap.

As a result, there will probably be some return from unwanted sources, which we will generally toss into the basket we call "noise." The goal is to minimize this noise and maximize the return. We can do this be assuming that the signal sent is the one received.

This assumption was more generally true in communications than it is in radar. In radar, we refer to a fictional "perfect reflector," which will reflect the pulse sent identically. Obviously, a complicated object such as a plane will

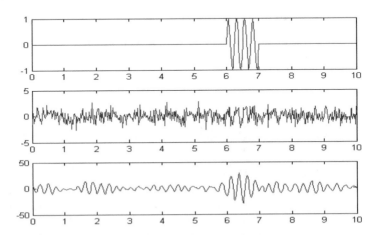

Figure 7.3: We illustrate using a matched filter above. The top signal is transmitted and received, but the noise produces the second graph, where the signal is obscured by the noise. The third graph shows that the noise can be suppressed, and the location or range of the target is essentially visible.

return the pulse but not perfectly reflected. The radar pulse will first bounce off of the nose, then the wings, then the tail, or continuously in some such manner. Thus, the return will be some collection of translates of the transmitted pulse. This is also a very clumsy assumption, but it does suit our purposes. Thus, we use convolution to look for translated versions of the pulse we transmitted.

7.3.2 Signal Averaging

A second tool for increasing the SNR is Signal Averaging. When we derived the Radar Range Equation 7.13 to describe the maximum possible range at which we could detect an object, the only free parameters to us seemed to be the antenna gain and the power. Unfortunately, the antenna gain is constrained by size, and on airborne radar systems cannot be changed much. The other problem is that the power can be increased, but when you increase it with an amplifier, generally the noise gets boosted also. Thus, these parameters can only be used to increase detectability to a certain defined limit.

Radar engineering has a definite advantage over communications engineering, however. Remember that in radar engineering we are collecting information literally at the speed of light. Thus, we can collect the information multiple times, in a very short period of time. If we were building a communications system, this would slow the bit rate, and as a result make the system slower. In radar, the constraints on speed are large, but generally accuracy is more important.

As a result, one can take the same measurements multiple times. The goal of this is that the true deterministic signal from a target should remain essentially unchanged for the extremely short time periods necessary. The random portion of the signal, created by randomness in the transmitter, receiver, environment, etc., should change. Random signals do not add coherently. Deterministic signals add coherently. We will now define these terms.

Definition 7.3.1 (Coherent and Incoherent Addition) *A deterministic gathered set of signals can be added coherently. This means that the sign of the signal will remain unchanged, and there will be no implicit subtraction in the signal. The signal will remain positive where positive, and negative where negative. A random signal will add incoherently. Thus, the sign of the signal will vary randomly, and the resulting sum will contain cancellation of the terms.*

Let us be more specific to our application. We are assuming that we have multiple returns of the type

$$r_k(t) = s(t) + n_k(t), \tag{7.14}$$

where $s(t)$ is the signal reflected off of the target, and $n_k(t)$ is random vector from $N(0, \sigma)$. Now if we average the terms $r_k(t)$, we get

$$\frac{1}{N} \sum_{k=1}^{N} r_k(t) = \frac{1}{N} \sum_{k=1}^{N} s(t) + \frac{1}{N} \sum_{k=1}^{N} n_k(t). \tag{7.15}$$

Now, the expectation of this random vector is certainly $s(t)$. The question is what is its variance. The variance will be the variance of the random vector, which will be

$$\sigma^2 \left(\frac{1}{N} \sum_{k=1}^{N} n_k(t) \right) = \frac{1}{N^2} \sum_{k=1}^{N} \sigma^2(n_k(t)) = \frac{1}{N^2} \sum_{k=1}^{N} \sigma^2 = \frac{\sigma^2}{N}. \tag{7.16}$$

Thus, coherent averaging can reduce the variance of the noise while maintaining the signal power. This is illustrated in Figure 7.4. There we increase the clarity of the target, by decreasing the effects of the noise.

There are certain limits and problems with this. The model in 7.14 which we used to show that this works, assumes that the return $s(t)$ does not vary with multiple acquisitions. Generally, the target and perhaps the radar (in airborne and other situations) are moving. On the other hand, the data is acquired at the speed of light, and the target and radar don't move that fast, thus is approximately true in most situations.

In addition, some of the interference may be coherent. Namely, there may be clutter, or problems with the machinery which are not random. This cannot be averaged away. Signal averaging remains a very fundamental tool, however.

7.3.3   Heterodyne Receivers

Another tool which is used in processing radar signals, after they are received by the antenna is the heterodyne. The reason for the heterodyne goes back to the basics, starting with the Shannon Sampling Theorem. If one sends a high frequency, then the return signal will be high frequency as well, and as a result, the bandwidth will be large. A large bandwidth demands very fast sampling and extra demands on the clock.

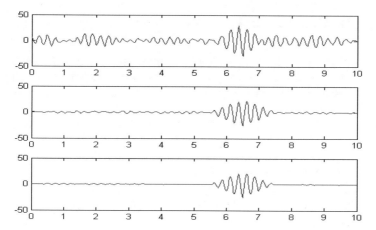

Figure 7.4: We illustrate using a matched filter with averaging above. Note that signals are gathered at the speed of light, and therefore gathering multiples of signals is often preferential to boosting the power. The top signal is transmitted and received and the matched filter is utilized as in Figure 7.3. The second graph was made using the same signal, but averaged 16 times with 16 different realizations of white Gaussian noise. The third graph features 128 realizations, or averaged signals plus noise.

In general, however, the frequency may be large, but the bandwidth may not be all that large. There are high frequencies, but essentially no low frequencies. Thus, we want to shift these "high frequencies" to intermediate frequencies (IF) where they can be sampled and manipulated more easily.

The mathematics is simple. Let us suppose that we have a signal which is

$$s(t) = w(t)e^{-if_k t}.$$

We want to shift it downward, so we multiply or mix this signal with another similar frequency and get

$$e^{if_m t}s(t) = w(t)e^{i(f_m - f_k)t}.$$

The resulting difference is a pulse with frequency $f_m - f_k$. Thus if we choose the frequencies appropriately, the "mixer," or heterodyne yields a received pulse which is downshifted in frequency so that it is much easier to handle.

Referring to Figure 7.5, we begin with a very high frequency signal, which is modulated by an exponential window. Its Fourier transform is displayed in the second graph. The third graph displayed that signal after mixing to reduce the frequency greatly. Thus, the fourth graph, which displays its Fourier Transform, illustrates the much lower bandwidth. The Shannon Sampling Theorem states that the last spectrum and final signal are much easier to sample than the first two.

7.3.4 IF Band-Pass Filters

The homodyne is utilized to shift the received signals to intermediate frequencies (IF), which are lower than the transmit/receive frequencies. The homodyne shift is utilized not only for sampling reasons, as outlined above. Generally, band-pass filters are available at the IF frequencies.

The reason for the band-pass filters is that the noise associated with the system and environment is usually widespread in frequency, or wideband. The signals are relatively narrowband. Thus, one can filter out the wideband noise which is not in the frequencies being utilized. The reason for doing this at IR is that there are generally physical, or analogue devices available which do this efficiently. Efficiency means that they get rid of the noise, and do not suppress the signal.

In Figures 7.6 and 7.7, we have illustrated the very basic process of IF band-pass filtering. The matched filter might theoretically eliminate the need for this process. The important difference is that this filtering is done before sampling, in actual hardware. Thus, much of the noise is eliminated before sampling concerns with noise become relevant.

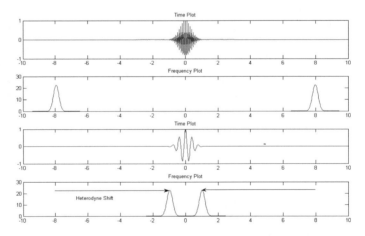

Figure 7.5: We illustrate a heterodyne shift to move the received signal to a lower IF frequency above. The top graph is the original signal. The second graph is the frequency graph of the original signal. The third graph is the signal after heterodyning the frequency down. The final graph shows the frequency shift caused by the heterodyne.

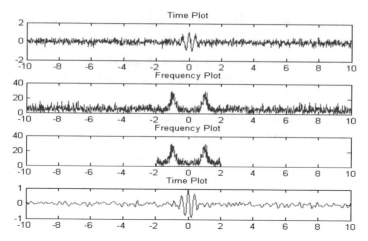

Figure 7.6: We illustrate the process of bandpass filtering above. The top graph is the signal embedded in noise. The second graph is the Fourier Transform of the top. The third is the filtered version of the second. The last graph is the time version of the filtered signal. Note the suppression of out of band noise.

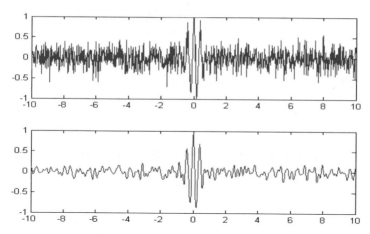

Figure 7.7: A direct comparison of the original noisy signal and the bandpass filtered signals are shown above. A good portion of the noise has been suppressed in the lower signal.

7.3.5   Problems and Exercises:

1. Use the $>> signal = [zeros(1, 512), sin([1 : 64]/64 * 4 * pi), zeros$ $(1, 512 - 64)]$; or something similar:

   (a) Take the FFT of the signal and observe where most of the power of the absolute value of this signal is.

   (b) Design a 0-1 linear filter in frequency which will preserve most 90% of this power.

(c) Now consider >> *nsignal = signal + randn(size(signal))*. Notice that the signal is almost undetectable.

(d) Consider the convolution of nsignal, and signal, using the command >> *newsignal = conv(signal, nsignal,' same')*. Is the signal now detectable?

(e) Figure out how to do the correlation of nsignal and signal. Is this better than the convolution? Why?

(f) Use the linear filter designed in (b) to do linear filtering via the FFT, the use the IFFT to return to the time domain. How does this compare to the convolution and/or correlation results?

(g) Write a brief program that will simulate signal averaging to reduce the noise. To do this you will have to add new noise each time. Begin by averaging 32 pulses, increase to 64, etc. How does signal averaging work compared to the other alternatives?

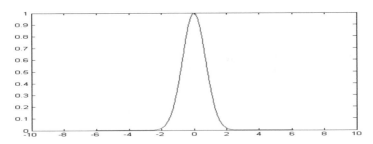

Figure 7.8: An example of the Point Spread Function from a radar system, or the return from one single target.

7.4 Revisiting Range Resolution: Uncertainty and Bandwidth

At this time, we need to revisit the issue of range resolution. We discussed it quickly in Section 7.1 but that was a very simple discussion which does not cover all of the issues. The issue which we want to discuss here is the effect of bandwidth on the range resolution. This is similar to our earlier discussion, but we will look in more detail.

The first question which we have to ask is how will a simple point target respond to our system. Suppose that we hang a bunch of 1-meter cylinders in the air (which has been done), and we want to know if we can separate them. The radar range equation gives us an idea of whether or not we can detect them. The second question is "How close can they be to each other before we can't tell them apart?". This is a question concerning uncertainty and bandwidth.

Forgetting noise, at this time, let us assume that the return signal is just a reflection of the signal which was sent, i.e.,

$$r(t) = s(t - 2R/c),$$

where $2R/c$ represents the time lag for a round trip.

The real question at this point is "What does the response of a single point target, after matched filtering, and other processing look like?"

Definition 7.4.1 (Point Spread Function) *The Point Spread Function of a remote system is its response to one remote point target, irregardless of noise.*

We begin with a single potential Point Spread Function, illustrated in Figure 7.8. Now let us suppose that we have two or three point targets, which are closely aligned, such as in Figure 7.9.

In the first graph of Figure 7.9, we see the three targets in three different colors. The reality is that they are not distinguished artificially by color, and the true return would be from the bottom graph. There they are indistinguishable.

The resolution problem in radar is very similar to the question of bandwidth and bit rate in communications. If one wants to separate bits in communications, then the old signal must die before the new one begins. This is exactly the same case in radar. As in communications, if you want the response to be quicker, the bandwidth must be larger. This is illustrated in Figure 7.10. With increased bandwidth in the transmitted pulse, or a shorter transmitted pulse, the targets can be distinguished.

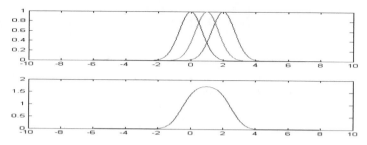

Figure 7.9: We illustrate three point targets, at three different ranges above. The first graph shows the individual responses coded by color. The receiver would get the second response. The second response is indistinguishable from a single larger target, and there is no individual coding of the targets.

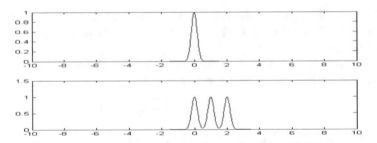

Figure 7.10: We illustrate three point targets, at three different ranges above. The first graph shows the individual response, but the receiver would get the second response. The second response is indistinguishable from a single larger target.

Let us now make this mathematical. We can do this from the equation which we initially described in Section 7.1, specifically 7.3.

$$\Delta R = \frac{c\Delta T_R}{2}. \tag{7.17}$$

In Section 7.1, we were primarily concerned about the speed of the clock, or the sampling rate. There are other issues here as we showed in Figures 7.9 and 7.10.

To address these issue, we must return to uncertainty and bandwidth. Namely, we must use the fundamental equation

$$\int t^2 |f(t)|^2 dt \int s^2 |\hat{f}(s)|^2 ds \geq \frac{1}{4}. \tag{7.18}$$

Loosely using Δ_t and Δ_f as the variance in time and frequency, this is rewritten as

$$\Delta_t \Delta_f \geq \frac{1}{4}, \tag{7.19}$$

or

$$\Delta_t \geq \frac{1}{4\Delta_f}. \tag{7.20}$$

Now, since size of the Point Spread Function is governed by the uncertainty principle, and we can see from Figures 7.9 and 7.10 that this is a limiting factor, we have that the time resolution

$$\Delta T_R \sim \Delta_t \geq \frac{1}{4\Delta_f},$$

or that

$$\Delta R = \frac{c\Delta T_R}{2} \sim \frac{c}{8\Delta_f}. \tag{7.21}$$

Thus, the range resolution doesn't only depend upon the clock. The range resolution is limited by the bandwidth. The larger the bandwidth, the higher the range resolution. Thus, let us memorialize the rules of range resolution.

Definition 7.4.2 (Range Resolution Rule) *The best way to obtain high range resolution is to transmit a short pulse, with a correspondingly high bandwidth.*

7.5   Doppler Measurement and Resolution

7.5.1   Doppler from a moving Target

We will now examine the use of doppler shifts in radar. We begin by studying the doppler effect of a wave off a moving target. In Section 7.1, we outlined some simple steps which can be made to estimate velocity. We also pointed out the inherent instability involved in indirectly measuring the velocity of an object through range measurements.

We will now explore measuring velocity directly through doppler measurements. Before we do this let us refresh our memories on basic terminology. First of all we know that the speed of light can be broken decomposed into the wavelength of a sinusoid, and its frequency, or $c = fw$, where f is the frequency and w is the wavelength. Thus, there are reciprocal relationships between the wavelength and the frequency, i.e., $w = c/f$.

Now, let us examine Figure 7.11. The reflective target, illustrated by the rectangle to the right, is moving at a speed of v_t. It makes contact with the leading portion of the sinusoid and continues to move to the left. The trailing edge of the sinusoid moves to the right at velocity c, so the trailing edge of the sinusoid and the target approach each other at a combined speed of $c + v_t$ (no individual pulse of object moves at more than the speed of light). Since distance divided by velocity gives time, we have that

$$t_d = \frac{w}{c + v_t}$$

is the time it takes for the trailing portion of the wave to reach the reflector. This is shown in the second graph. The leading edge of the reflected sinusoid continued to the left at the speed of light. It did not return to the "starting point," however, since the time until the trailing edge met with the reflector was reduced by the velocity of the reflector.

The wavelength of the resulting "compressed" wave is therefore $w_d = w - d_1 - d_2$. Notice, however, that $d_1 = d_2$, since the time that the trailing edge would have needed to reach the original location of the reflector, on the right, is exactly the time it would have taken the leading edge to "return" to the original point at the left. Thus, we have $w_d = w - 2d_1 = w - 2v_t t_d$, or

$$w_d = w - 2\frac{wv_t}{c + v_t} = w\left(1 - \frac{2v_t}{c + v_t}\right) = w\frac{c - c_t}{c + v_t}. \tag{7.22}$$

Since $w = c/f$, we have

$$c/f_d = c/f \frac{c - v_t}{c + v_t}$$

or

$$f = f_d \frac{c - v_t}{c + v_t}.$$

Rearranging, we get

$$f_d = f \frac{c + v_t}{c - v_t}.$$

Returning to 7.22, we can substitute immediately for w_d and w to get

$$c/f_d = c/f \left(1 - \frac{2v_t}{c + v_t} \right),$$

or

$$f = f_d \left(1 - \frac{2v_t}{c + v_t} \right) = f_d - f_d \frac{2v_t}{c + v_t} = f_d - f_s. \qquad (7.23)$$

Thus, we have the new doppler frequency is $f_d = f + f_s$, where we refer to

$$f_s = f_d \frac{2v_t}{c + v_t} = f \frac{2v_t}{c - v_t} \qquad (7.24)$$

as the doppler shift frequency.

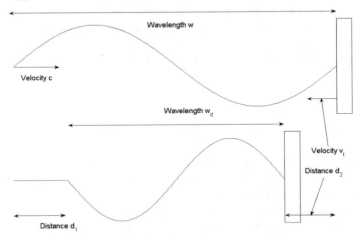

Figure 7.11: We illustrate a basic doppler shift above. At the right is the target, represented by the rectangle, striking the leading edge of a single wavelength sinusoid. The sinusoid is moving at the speed of light, and the target at velocity v_t, so they are moving toward each other at a speed of $c + v_t$. The second graph reflects the reflected wave. Note that the reflected wave is compressed.

7.5.2 Doppler from a moving Source

There are slight modifications which have to be made when the source is moving. These are highlighted in Figure 7.12. In this case the source, at the right, is moving to the left at velocity of the emitter v_e.

In Figure 7.12, we see the top graph, where is source is not moving. The bottom graph has a velocity of v_t, and therefore, the source moves while the wave is emitted. Since the leading edge of the wave still only travels at the speed of light, the leading edges from both the moving and non-moving sources are in the identical place. Thus, the wavelength of the moving target changes. The emission time is simply $t_e = w/c$. Thus, the distance moved is $d = t_e v_e$. The new wavelength is thus

$$w_d = w - t_e v_e = w - (w/c)v_e = w\left(1 - \frac{v_e}{c}\right).$$

We once again use the fact that $w = c/f$, and get $c/f_d = c/f(1 - (v_e)/c)$ or $f = f_d(1 - (v_e)/c)$. Solving for f_d, we get

$$f_d = f + f_d\frac{v_e}{c} = f + f_s, \qquad (7.25)$$

which implicitly defines our doppler shift f_s. Solving more specifically for f_d yields

$$f_d = \frac{c}{c - v_e}f,$$

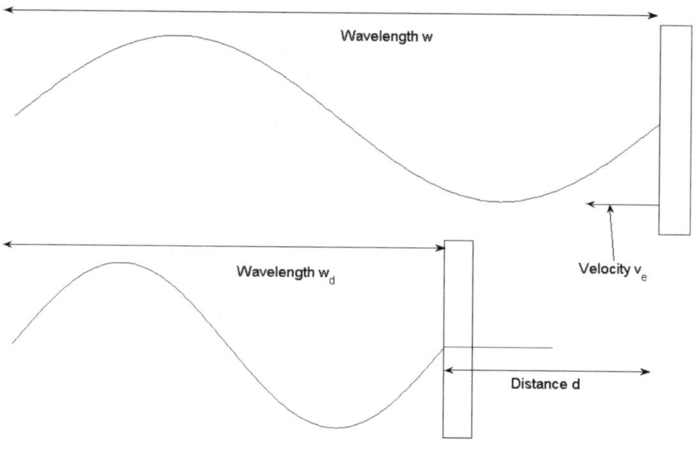

Figure 7.12: We illustrate a doppler shift from a moving source above. At the right is the source, represented by the rectangle. If the velocity is zero, as in the case of the top graph, then the wave is emitted in time $t_e = w/c$. If the velocity is non-zero, or v_e as in the bottom diagram, then the source still emits the wave in time $t_e = w/c$. The only difference is the source have moved a distance $d = v_e t_e$. The wave is obviously compressed, by a factor of d, or the wavelength $w_d = w - d$.

or that

$$f_s = \frac{c}{c - v_e} f \frac{v_e}{c} = f \frac{v_e}{c - v_e}. \tag{7.26}$$

One thing stands out quite clearly here. The doppler shift of the emitter 7.26 is exactly 1/2 of that of a moving target 7.24,

$$f_s = f \frac{2v_t}{c - v_t}.$$

This is quite interesting, but at the same time makes some sense. Examining Figures 7.12, and 7.11 gives one some insight into the phenomena.

7.5.3   Doppler Resolution

We must now examine the necessary conditions for resolving or detecting a doppler shift. Regardless of whether the shift is from a target or source, we will assume that there is a non-zeros doppler shift f_s. The question is whether or not we can detect it. We will once again assume that we are dealing with a point target.

For range resolution, we wanted to be able to accurately measure the range, or resolve the difference between different targets at different ranges. Since range is naturally a time consideration, with different echos returning at different times, this was primarily an issue of the makeup of the pulse in time.

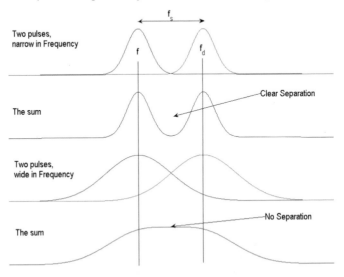

Figure 7.13:   We illustrate the problem of resolving a doppler shift above. In the first graph, the frequency content of two pulses, one doppler shifted and one not, are illustrated. The second graph shows that the doppler shift is still detectable, even when the pulses are added. The third graph shows two pulses, as the first, but with much wider frequency content. The fourth graph shows their sum, and illustrates that this doppler shift is undetectable.

Doppler, however, is quite obviously an issue of detecting the doppler shift f_s, which is a frequency consideration. To develop the doppler shifts, we utilized single cycles of a pure sinusoid. That is not generally realistic for a deployed system. Let us assume for now that we are sending some type of pulse, with a very distinct center frequency f. Linearity assures us that the doppler shift will be the same (you break down the pulse into its separate frequencies).

We illustrate this in Figure 7.13. The idea is that there are two point targets at the same range, one which is moving and has a doppler shift of f_s. The other is not moving. The question is "Can you detect the difference between these two?". The first graph in Figure 7.13 illustrates the frequency spectrum of two pulses, which have relatively tight frequency representations. The second graph shows the return, where the two spectra are added. The frequency shift f_s is still quite visible, and thus detectable.

The third graph in Figure 7.13 shows the frequency content of two pulses which are more widely spread in frequency. The fourth graph shows that when these pulses are added, the frequency shift is not detectable. Thus, the targets are not easily resolvable.

Now remember that because of the uncertainty principle, getting a pulse which is very tightly spread in frequency means that it must be longer in time. Indeed in Figure 7.13, the only differences between the first example and the second was that we dilated the pulses. The simple answer is that making your pulses longer in time, makes them more tightly concentrated in frequency. Thus, we have the following.

Definition 7.5.1 (Doppler Resolution Rule) *The best way to obtain high doppler resolution is to transmit a long pulse, with a correspondingly small bandwidth.*

You might remember that this rule is quite different than the earlier rule we had for range resolution. For clarity, we will state them in one place.

Definition 7.5.2 (The Resolution Rules) *We will now state our two, opposing Rules.*

- **The Range Resolution Rule** *The best way to obtain high range resolution is to transmit a short pulse, with a correspondingly high bandwidth.*

- **The Doppler Resolution Rule** *The best way to obtain high doppler resolution is to transmit a long pulse, with a correspondingly small bandwidth.*

Obviously, these two rules are not compatible with each other. It is often necessary to design a radar system with high range resolution and high doppler resolution. The simple case of the interstate highway gives an example. When measuring velocity, or doppler, it is preferential to know which vehicle is creating that signal. Thus, you need to accurately know the speed of the target, and which target it going at that speed. The fact that this is difficult is a classical uncertainty problem.

These questions have been at the center of radar design for the last 80 years.

7.5.4 Problems and Exercises:

1. **Design of Doppler Pulses:** We want to design a pulse which is optimized for detecting the doppler shift of a moving vehicle or target. To do this, we need a pulse which is short enough to separate multiple targets in range and well enough localized in frequency to accurately detect the doppler shift. Consider an arbitrary pulse beginning with a signal of the type

   $$>> doppler = [zeros(1, 256), sin([1 : 512]/512 * 2 * pi * 64). * w, zeros(1, 256)];$$

   where w is a windowing function on 512 pixels designed to localize the doppler pulse in frequency. Consider several windowing functions w.

   (a) Simply let $w = 1$, in other words not really a window whatsoever.

   (b) Choose w to be a truncated Gaussian which has its center in the middle of the 512 pixels and is essentially zero at the ends of the window.

   (c) Choose w to be a raised cosine window as discussed in section 5.2.3.

   (d) Perhaps another smooth window of your own design, which meets the criterion of section 5.2.3.

   Now, analyze the 3 or 4 window choices above for localization in frequency in the following ways: (i) By plotting the absolute values of their FFTs. Visually decide which is best, or if there is a winner. (ii) By somehow measuring the concentration of energy around the central peak of their FFTs. Explain how you analyzed this, and if there is a winner, and if it is the same winner.

7.5.5 Chapter Project:

Most of the exercises in this Chapter are extensive and could be projects on their own. Depending on time constraints, one could include all of them in a Chapter Project, or a select subset.

Chapter 8
Image Processing

We have studied the use of the Fourier Transform on one-dimensional problems on a number of problems. Fourier Analysis is also extremely valuable as a tool to understand multidimensional problems. Modern cable systems utilize Fourier Analysis to compress video signals in order to provide more channels through the same connections. The military uses Fourier Analysis to analyze images and detect potential targets of interest. There is no shortage of applications, so we will now move on with the formalities of the multidimensional Fourier Transform.

8.1 Fourier Analysis in $\mathbb{R}^n$

Just as with functions on the real line, or $\mathbb{R}$, the Fourier Transform for functions on $\mathbb{R}^n$ will come in three varieties: (1) Fourier Series on $L^2(D)$, where D is a domain such as $[-\pi, \pi] \times [-\pi, \pi]$; (2) The discrete Fourier Transform on $m \times n$ arrays; (3) The Fourier Transform on $L^2(\mathbb{R}^n)$. We will not treat these individually as we did in the first three chapters. All of the theorems and results of those chapters have analogies. The interested student should try to prove them! The proofs are nearly identical.

We will instead primarily concentrate on $L^2(\mathbb{R}^2)$ and point out problems that may come up when dealing with the other situations. Let us begin with Fourier Series on $[-\pi, \pi] \times [-\pi, \pi]$. We refer to a square domain such as $[-T, T] \times [-T, T]$ as D_T. In an analogue with the 1-D situation, we have

Theorem 8.1.1 (Two-Dimensional Fourier Series) *Let $f(x, y) \in L^2(D_\pi)$. Then we have the representation*

© Springer Science+Business Media, LLC 2017
T. Olson, *Applied Fourier Analysis*,
https://doi.org/10.1007/978-1-4939-7393-4_8

$$f(x,y) = \frac{1}{2\pi} \sum_m \sum_n c_{m,n} e^{-i(mx+ny)},$$

where

$$c_{m,n} = \int_{-\pi}^{\pi} \int_{-\pi}^{\pi} f(x,y) e^{i(mx+ny)}.$$

Note that this representation will be periodic in both x and y. Thus, it is only valid on the stated domain.

We calculated several Fourier Series in one dimension. Similar series can be calculated in two dimensions, but the effort necessary becomes tedious. We will primarily concentrate on understanding. Therefore, we introduce the continuous Fourier Transform.

Theorem 8.1.2 (Two-Dimensional Fourier Transform) *Let $f(x,y) \in L^2(\mathbb{R}^2)$. We define the two-dimensional Fourier Transform to be*

$$\hat{f}(u,v) = \frac{1}{2\pi} \int \int f(x,y) e^{i(xu+yv)} \, dx \, dy.$$

In addition, we have an inversion formula

$$f(x,y) = \frac{1}{2\pi} \int \int \hat{f}(u,v) e^{-i(xu+yv)} \, du \, dv.$$

The proofs of these theorems are analogous to the one-dimensional proofs and can be found elsewhere.

The question which we want to ask now is "What does a component of the two-dimensional Fourier Series, or Fourier integral look like?" In one dimension, we had a clear idea of what a sine or cosine of a certain frequency looked like. "What do the component functions look like in two dimensions?" The answer is relatively easy.

A basic function in a Fourier Series is $\cos(mx+ny)+i\sin(mx+ny)$. Let us concentrate on $\cos(mx+ny)$. This looks a lot like a one-dimensional function, and in a sense it is. It is a one-dimensional wave propagating across the two-dimensional plane. This can be understood by looking at the "level sets" of the function or the places where $\cos(mx + ny) = k$. These are clearly places where $mx+ny = k_2$. Thus, the level sets are lines. These sets of parallel lines are the level sets on the wave, which is oriented in a specific direction. We illustrate this with $m = 5$ and $n = 7$ in Figure 8.1.

We understand that the Fourier Transform in two dimensions is built from plane waves. Understand that the slope of the level curves, which can be solved for as $y = -m/nx + k_2/n$ or $-m/n$, determines the orientation of the plane wave. Thus, the plane waves give us direction and frequency. This is the nature of the 2-D transform.

8.2 Edge Detection

We will now consider one common technique used to analyze images, called edge detection. The purpose is to find rapid changes, so that separate objects can be identified. In manufacturing, you might want a robot to recognize a certain bolt or screw. Its outline, or edge profile, will generally give a definitive clue as to where it is. From a military standpoint, tanks or artillery pieces must be located from vast amounts of satellite data. Their edge profiles are also a definite indicator of where they are. In addition, background, such as foliage generally, has a very different edge profile.

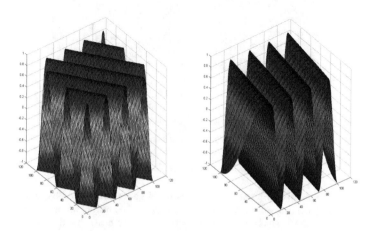

Figure 8.1: The nature of a plane waves is illustrated with two examples above. The two-dimensional Fourier Transform is built from plane waves, of various frequencies, and moving in various directions. Thus, there are effectively two variables in the 2-D transform, direction, and frequency.

What is an edge? If you are looking at a landscape, it can be the edge of a forest, a cliff, or a river. A basic definition is a rapid change. Thus, if we are going to look for rapid changes, we want to look at derivatives. Since we are currently considering two dimensional images, we want to consider partial derivatives. Specifically, we want

$$\frac{\partial}{\partial x} f(x, y) = \lim_{h \to 0} \frac{f(x + h, y) - f(x, y)}{h} \tag{8.1}$$

and

$$\frac{\partial}{\partial y} f(x, y) = \lim_{h \to 0} \frac{f(x, y + h) - f(x, y)}{h}. \tag{8.2}$$

We often refer to the partial with respect to x as f_x and the partial with respect to y as f_y.

Dealing with Color, RGB, or .jpg images in MATLAB

Usually, we utilize color images, although black and white images can be found. The easiest place to get images is from the Internet. Usually, they are easily downloadable with a few clicks of your mouse.

Color images are generally generated in rgb format. That is, they have one variable for red, one for green, and one for blue. Thus, if you have a 512×512 image (rarely are they sized like this), you will have 3 512×512 files in that image. Thus, you have a three-dimensional array, with three colors for a two-dimensional pictures.

Figure 8.2: The nature of a plane wave is illustrated above. The two-dimensional Fourier Transform is built from plane waves, of various frequencies, and moving in various directions. Thus, there are effectively two variable in the 2-D transform, direction, and frequency.

We will generally only operate on each of these three images individually and then perhaps combine the result. For edge detection, it seems that most you can usually find edges with any of the color arrays, because most objects have each of the three base colors in their description.

8.2.1 Horizontal and Vertical Edges: Standard Approximations

We first calculate the approximate partial derivatives with respect to x picture shown in Figure 8.2. We use the two-point approximation and only the first of the three color arrays. We then use all three of the color arrays, add the three approximate derivatives, and use that as our approximation. This seems qualitatively to be somewhat better, so we show that in the top picture of Figure 8.3.

Secondly, we take the y partial derivatives, as described above.

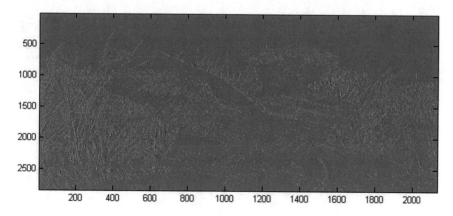

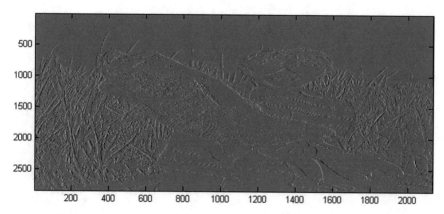

Figure 8.3: Above, we calculate the partial derivative with respect to x in two ways. First by using a two-point approximation as is normally used for the theoretical definition, which is shown at the top. Secondly, we took larger averages, to the left, and the right, and took their difference. This is shown on the bottom.

Recall from multivariable calculus that the gradient is defined by $\nabla f = (f_x, f_y)$.

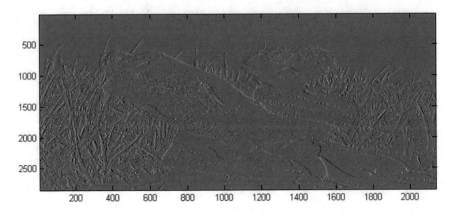

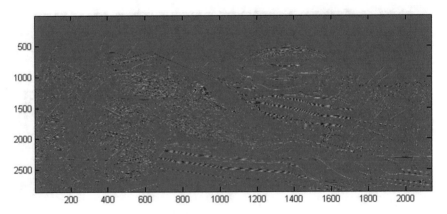

Figure 8.4: We contrasted the vertical edges, above, against the horizontal edges below. Notice that most of the swamp grass, or cattails, is vertical and disappear. The detail on the sunglasses changes, indicating vertical and horizontal features. Many edges, however, are not either horizontal, or vertical.

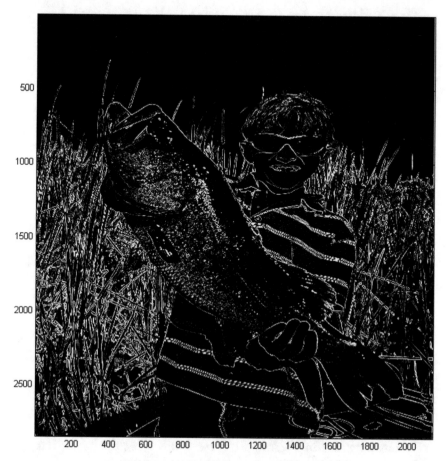

Figure 8.5:   We illustrate the absolute value of the gradient above. This is $|\bigtriangledown f|^2 = f_x^2 + f_y^2$. Note that this does pick up most of the edge detail of the picture.

8.2.2 Exact Partial Derivatives and Directional Derivatives

We used the standard approximations, 8.1, 8.2 above. We can also calculate partial derivatives directly from the Fourier representation

$$f(x,y) = \frac{1}{2\pi} \sum_m \sum_n c_{m,n} e^{-i(mx+ny)},$$

where

$$c_{m,n} = \int_{-\pi}^{\pi} \int_{-\pi}^{\pi} f(x,y) e^{i(mx+ny)}.$$

Calculating directly, we have

$$f_x(x,y) = \frac{\partial}{\partial x} f(x,y) = \frac{\partial}{\partial x} \frac{1}{2\pi} \sum_m \sum_n c_{m,n} e^{-i(mx+ny)},$$

$$= \frac{1}{2\pi} \sum_m \sum_n \frac{\partial}{\partial x} c_{m,n} e^{-i(mx+ny)}$$

$$= \frac{1}{2\pi} \sum_m \sum_n (-im) c_{m,n} e^{-i(mx+ny)}. \tag{8.3}$$

Thus, we multiply the coefficients $c_{m,n}$ by $-im$ to get the partial. Similarly, we multiply by $-in$ to get the partial with respect to y. Now, we must be careful to multiply by the appropriate coefficients.

We illustrate this in Figure 8.6. There is not a big difference, but this might become more of a problem if there were noise present.

8.3 Interpolation: High-Definition TV and beyond

We will now consider the problem of taking a low-resolution image and trying to make a high-resolution image with favorable visual characteristics. What we don't want to do is have he image look blocky, or jumpy, or broken. Rather, we want the same original pixels, with the additional pixels coming from a smooth and reasonable interpolation. This is exactly the problem of taking an old, low-resolution picture and making a high-resolution picture suitable for a modern TV.

8.3.1 Revisiting Shannon and Advanced Interpolation

Assuming that we have a fixed number of pixels, we might as well assume we have the values of $f(x,y)$ on the integers (m,n), or we have the values $f(m,n)$. Thus, we want an analogue of the one-dimensional Shannon Theorem which is

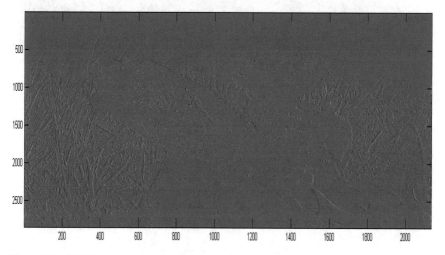

Figure 8.6: We illustrate taking partials according to the "exact" derivative formula above. They are very similar. Real differences may become apparent if the image is noisier.

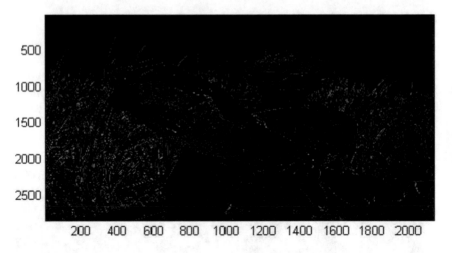

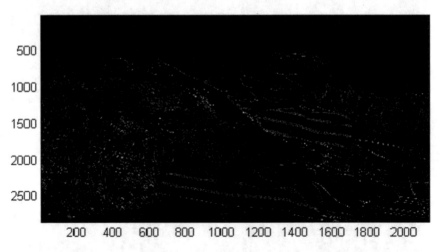

Figure 8.7: We illustrate the differences between directional partial derivatives above. These partials are taken at approximately $\pi/6$ and $4\pi/6$. You can clearly see the stripes on the boy's shirt in one, and not on the other. Note that some of the background has partials in all directions, and the difference between partials is not obvious.

Theorem 8.3.1 *Assume that $\hat{f}(u,v) = 0$ for $|u| > \pi$, and for $|v| > \pi$. Then we can write*

$$f(x,y) = \sum_{m}\sum_{n} f(m,n)\frac{\sin(\pi(x-m))}{\pi(x-m)}\frac{\sin(\pi(y-n))}{\pi(y-n)}.$$

Figure 8.8: The following image was submitted to a local newspaper, for publication. It had to be enhanced to meet the necessary requirements for publication.

While the above formula can be used, it is slow, and we generally do not use it. Instead, as in the 1-D case, we use the 2-D DFT, or FFT, to calculate the Discrete Fourier Transform of the discrete data $f(m,n)$. Once this is calculated, then we embed this Fourier Transform in a larger array of zeros, implicitly making the higher frequencies zero. An Inverse Fourier Transform then completes the task. When we did interpolation in the 1-D case, however, we only had one level of data.

With images, we generally have three-level data. Namely, the data is RGB (Red, Green, Blue), giving the colored image that we see. This is stored as a three-dimensional array, generally nxmx3. This is most easily taken care of with an example.

We begin with a typical low-resolution image, shown in Figure 8.8. This is a $700 \times 700 \times 3$ image. This image was submitted to a paper, but it was necessary to increase its resolution for publication. We want to increase the resolution to 1400x1400. Let us go through the necessary steps at this time.

The first step is to calculate its Fourier Transform. We do not calculate the Fourier Transform of the entire 3-D array, but rather we calculate the Fourier Transform of each of the colors, individually.

Figure 8.9: We illustrate the very simple process of interpolation above. The fft of each of the rgb bands is taken first. Then, each of the three Fourier Transforms is embedded in an array of zeros which is twice its size. These embedded arrays are then Inverse Fourier Transformed, and the final result is an image which is twice as big.

8.3.2   Regaining a .jpg image in MATLAB: Normalization and Format

A small but important step of the above process is to allow the image to be restructured so that it may be saved once again as a .jpg image. The key to this is to understand that the .jpg image is stored in an 8-bit format; i.e., the values are from 0 to 255. In MATLAB, this is called unsigned integer 8, or uint8. Secondly, the normalization must be correct. In the case above, it was necessary to multiply by 4, in order to renormalize the image. If the image were enlarged further, a larger multiplier would have to be used. The final step in MATLAB is to use the command uint8, to convert the interpolated and renormalized image to the proper format.

8.4 Compression

One of the goals of Fourier Analysis is to be able to represent a function, or image, in a very efficient way. We have talked about decay rates, and how cosine and sine expansions can be utilized in order to efficiently store even or odd functions in one dimension. We would like to be able to efficiently store images, primarily for the decreased cost of transmission (if you are a cable TV company).

We will begin with a basic approach to compression, which we discussed in the last section. We will then track the error mathematically and perhaps visually. Visual perception of the image is very difficult to track and is a very active study area. We will then look into more advanced methods for compression.

8.4.1 Shannon Sampling, Compression, and Interpolation

We introduced the process of Shannon interpolation in the last section 8.3.1. The idea here is to keep a subsampled version of the image and then use Fourier interpolation, or the Shannon Sampling Theorem to reconstruct the image. We will then compare this idea to other methods of compression. Four images with the same number of coefficients are compared in Figure 8.10. For clarity, we now outline the algorithm.

Shannon sampling and compression

1. Subsample the image on each of the 3 rgb layers by a factor of k. Thus, you want to keep every k^{th} pixel from the image.

Shannon decompression or interpolation

1. Take the Fourier Transform of each rgb layer.

2. Embed the subsequent shifted Fourier Transforms in an array of zeros, which has the size of your desired final image.

3. Inverse Fourier Transform, the individual rgb layers.

4. Convert the final results into unsigned integer 8 formats (uint8 in MATLAB).

In Figure 8.10, we subsample the image in space, initially keeping every 16th coefficient for a $512/16 \times 512/16 = 32 \times 32$ image. These have to be sampled for each layer of the rgb image. We then interpolate with Shannon interpolated described earlier. The first image had a total L^2 error of more than 10 percent. The second image, which kept 64×64 pixels and then interpolated, had a relative error of 6 percent. The third with 128×128 pixels kept had an error of 4 percent. These images seem somewhat unacceptable. This is because we have violated the rules of the Shannon Sampling Theorem, and therefore, we have undersampling and aliasing. The third image 8.10, however, seems close original in 8.10. Once again, mathematical measures may not accurately predict what we see with our eyes.

8.4.2   Fourier Truncation

We now test an even simpler idea. The idea is purely to keep the lowest Fourier frequencies and test the accuracy of the algorithm. This may seem identical to Shannon sampling and interpolation, but it is not. We will "sample" the lowest frequencies of the Fourier Transform. Keep them and then restore an approximation from these coefficients. You must remember to convert the image back to unsigned integer 8 (uint8) format in MATLAB to have it displayed correctly and to properly store it as a .jpg image which can be read and displayed by common Web browsers and image processing packages. For clarity, we once again detail the process.

Fourier compression

1. Take the Fourier Transform of each of the rgb layers.

2. Keep only the lowest Fourier shifted coefficients (in MATLAB use fftshift).

Fourier decompression and interpolation

1. Embed the individual Fourier Transforms in arrays of an arbitrary size of your liking.

2. Do a proper inverse FFT on each of the rbg layers.

3. Make sure the result is real and convert to unsigned integer 8 (uint8 in MATLAB).

Figure 8.10: We illustrate Shannon sampling, compression, and interpolation technique above. We subsample the pixels, i.e., keeping every 8th pixel from the original image, and then perform the Fourier Transform on this subsampled version. A very blurred image using Shannon compression and interpolation is shown at top right with a compression factor of 64. At the bottom right, we see the Shannon compressed image at a rate of 16. At bottom left, we see the result of using Shannon compression with a factor of 4. The relative errors are .10, .06, and .35, respectively. These are very close to the numbers for periodized Fourier truncation, although visually Fourier truncation seems better.

Counting the coefficients correctly

We must remember that we started out with real coefficients. The Fourier coefficients are complex, requiring that we keep a real and complex coefficient at each frequency point. Thus, we must be careful how we count.

We begin by keeping the lowest 64×64 coefficients of the 2-D transform. Because of the above discussion, we realize that we can only keep the coefficients of the lowest 32 frequencies, since each frequency coefficient is actually two coefficients. We follow with 128×128 and 256×256 coefficients, respectively. Since the original image was 512×512, these represent compression rates of 64, 16, and 4, respectively. Note that the first image which is blurred has an L^2 relative error of .10, the second image is also blurred .07, and the third, which is acceptable but somewhat blurred, has an error of .045. Thus, the error seems small, but the detailed portions of the image, which are necessary for the images and recognition of features, are small in terms of the total energy of the image. We must therefore be careful to make sure we recognize what is significant when doing compression. The L^2 error gives us some idea, but is not a great measure of what the eye sees.

8.4.3   Periodization and Fourier Truncation

One of the problems with truncating an image is the artificial creation of discontinuities at the edges of the blocks. This keeps the Fourier Series or the DCT from converging quickly. Realize that the 2-D Fourier Series is by its very nature periodic both vertically and horizontally. Thus, when we truncate the Fourier Series, the tendency is for the left and right edges to try to come to a continuous agreement and the top and bottom edges to come into continuous agreement. This causes Gibbs ringing which is clearly visible in the images of Figure 8.11.

One way to combat this problem is to try and periodize an image. Let us imagine an image as a function in the first quadrant, with x and y constrained between 0 and π. The image will almost never have its top and bottom be equal, nor will its left and right edges be equal. Thus, we would like to make a new image, with the old image embedded in it, whose periodic extension will be 2π periodic in both variables. In addition, this new image will not have discontinuities induced at any of the edges.

To understand this, let $f(x, y)$ be a simple function of the variables x and y which range from 0 to π. Let us now define an extension of this function $f_{ext}(x, y)$ where $-\pi \leq x, y \leq \pi$ in the following way. Imagine flipping the image in the first quadrant about the y-axis. This defines f_{ext} in the second quadrant. Mathematically, we have $f_{ext}(x, y) = f(-x, y)$ for $-\pi \leq x \leq 0$ and $0 \leq y \leq \pi$. Notice that this newly defined function is continuous at the boundary created by the positive y-axis. Next, visually, flip the image in the second quadrant about the x-axis to give the extension in the third quadrant. Thus, $f_{ext}(x, y) = f(-x, -y)$ for $-\pi \leq x \leq 0$ and $-\pi \leq y \leq 0$. Once again, continuity is maintained along the negative x-axis. Finally, we flip the image in the third quadrant across the negative y-axis, or $f_{ext}(x, y) = f(-x, -y)$ for $0 \leq x \leq \pi$ and $-\pi \leq y \leq 0$. We will call this the **four flop method**.

Figure 8.11: We illustrate a the Fourier compression algorithm above. At top left, we show the original image. A very blurred image created form keeping the lowest 32×32 complex coefficients from the Fourier Transform is displayed at the top right. Note that this represents compressing the image by $64^2/512^2 = 64$, but the result is blurred. At the bottom right, we see an image compressed to 32×32 complex coefficients, resulting in a compression of 16. This is still blurry, but a slightly more reasonable image. Finally, we keep 64×64 complex coefficients, with a compression factor of 4. This image is reasonable but displays ringing at the edges and some degradation of the small features.

Notice that we have continuity everywhere across the axes. Moreover, the value of f_{ext} along the top border $y = \pi$ is exactly the value of f_{ext} along the bottom border $y = -\pi$. The same is true along the borders $x = \pm\pi$. Thus, if $f(x, y)$ was continuous in the first quadrant, the extension $f_{ext}(x, y)$ is also continuous. Furthermore, the periodic extension of $f_{ext}(x, y)$ will be continuous on the whole plane. As we discussed in Chapter 2, inducing discontinuities at the endpoints of an interval ensures that the Fourier Series will not converge quickly. This is also true in 2-D. We have now avoided this problem. For clarity, we explicitly outline the algorithm.

Counting the coefficients correctly

When we utilized the Fourier Transform on an arbitrary image, the coefficients were both real and complex. Because we have made the image even in all coordinates, the Fourier Transform of our periodized image will only have cosine terms or will be purely real. Therefore, we can take the lowest 32×32 real coefficients of the Fourier Transform and still have the appropriate compression ratio.

Fourier periodization and compression

1. Use the four flop method to periodize the image

2. Take the Fourier Transform of each of the rgb layers.

3. Keep only the lowest Fourier shifted coefficients (in MATLAB use fftshift).

4. Recognize that you only have cosine coefficients, don't store the sine (complex) coefficients.

Fourier decompression and interpolation

1. Embed the individual Fourier Transforms in arrays of an arbitrary size of your liking.

2. Do a proper inverse FFT on each of the rbg layers.

3. Make sure the result is real and convert to unsigned integer 8 (uint8 in MATLAB).

8.4.4 Blocking, Compressing, and Smoothing Images

We will now try some aggressive techniques for compression. Recall that the Fourier Transform is not very efficient at representing any portion of a 1-D or 2-D function, such as an image if there are discontinuities in that image. Since nearly all images have edges within them, it is unrealistic to think that trying to use the Fourier Transform on the whole image will be the most efficient way to store the image.

Figure 8.12: We illustrate the periodization of a subportion of the bee image above. We begin with the first quadrant, upper right. We then create a mirror image of this by flipping the first quadrant about the positive y-axis to get the image in the second quadrant. The third quadrant is created by flipping the second quadrant about the negative x-axis. Finally, the fourth quadrant can be completed by flipping the third quadrant about the negative y-axis, or the first quadrant about the positive x-axis. The result is an image with no interior edge discontinuities and no edge discontinuities between the top and bottom, or left and right edges.

Figure 8.13: We now utilize Fourier truncation as in Figure 8.11 but on the periodized image. We also keep 64^2, 128^2, and 256^2 coefficients for compression ratios of 64, 16, and 4, respectively. These are shown in the above image, with the most compressed at top right, the middle compressed at bottom right, and the least compressed (factor of 4) at bottom left. Realize that these coefficient come from an artificially large array created in Figure 8.12. The relative errors were, .09, .06, and .036, respectively. This appears to outperform basic Fourier truncation. It is unclear, however, whether L^2 error is truly an accurate predictor of image accuracy. The human eye is fickle with images.

Figure 8.14: We illustrate a comparison of direct Fourier truncation and Fourier truncation of a periodized image above. The periodized image at a compression rate of 4 is at the bottom left. The Fourier truncated image with a compression rate of 4 is at bottom right. For comparison, the periodized compression at a rate of 16 is at top left and the non-periodized image with a compression ratio of 16 is at top right. The periodized images do not show as much Gibbs' ringing and seem to outperform the unperiodized images.

Thus, we will now consider trying to represent an image locally with the Fourier Transform. Many images have regions which are relatively constant, or without discontinuities or edges. One might think of any outside image with a blue-sky background. While the foreground may have a lot of detail, the background can be very efficiently represented.

We will arbitrarily choose to look at 64×64 sub-blocks of our image and then trying to compress them. This may be more efficient, but does cause some additional problems. We utilize the Fourier truncation method, i.e., discarding the higher-frequency components described above, on each of the 64×64 blocks. Remember that we also need to do this for each of the rgb (red, green, blue) layers. The first result of utilizing this approach is shown in Figure 8.15, at the top right. We have compressed this by a factor of 16. The problem we see with this composite image is the discontinuities at the edges of the blocks.

There are many approaches to fixing the discontinuities at the edges. We chose for simplicity to throw away the high-frequency components of the composite image. This produces the more visually appealing image at the bottom left of Figure 8.15. This might seem to change the compression of the image. Indeed, it would, but we can store the compressed image and then smooth when we reconstruct the image, without changing the compression ratio. The percentage error is similar to earlier versions, but the final question is "How good does the image look?". This looks reasonable with the high compression ratio of 16.

The final image at the bottom left of Figure 8.15 is done the same way, with a compression ratio of 4. This image seems very good. We will compare the best of all of these methods in a later section. Once again, for clarity, let us outline the steps.

Fourier blocking and compression

1. Pick an appropriate number to subdivide (block) the image into (blocking).

2. Take the Fourier Transform of each block of the rgb layers.

3. Keep only the lowest Fourier shifted coefficients (in MATLAB use fftshift).

4. Realize that you have complex coefficients in your compression statistics.

Fourier decompression and interpolation

1. Embed the individual Fourier Transforms in arrays of an arbitrary size of your liking.

2. Do a proper inverse FFT on each of the rbg layers.

3. Make sure the result is real and convert to unsigned integer 8 (uint8 in MATLAB).

Figure 8.15: The test Hungry Bee image is shown at the top left. The image at the top left is the blocked and compressed image, which shows some border problems. With a compression ratio of 16, however, this is not too bad. The image at the bottom right is the image at the top right, with the high Fourier frequencies removed. This is much more pleasing, but actually has a higher percentage error. The image at bottom left shows the same process, but with compression of only 4. This image is nearly perfect.

8.4.5 Periodization and Blocking

We have discussed earlier that periodization would reduce the necessary number of Fourier coefficients. Blocking actually increases the need for periodization, since the individual blocks induce massive edge effects, as we have seen in Figure 8.15. The obvious step we should take is to periodize the blocks and therefore minimize the edge effects.

Fourier blocking, periodization, and compression

1. Pick an appropriate number to subdivide (block) the image into (blocking).

2. Use the four flop method to periodize the sub-blocks of image

3. Take the Fourier Transform of each block of the rgb layers.

4. Keep only the lowest Fourier shifted coefficients (in MATLAB use fftshift).

5. Recognize that you only have cosine coefficients, don't store the sine (complex) coefficients.

Fourier decompression and interpolation

1. Embed the individual Fourier Transforms in arrays of an arbitrary size of your liking.

2. Do a proper inverse FFT on each of the rbg layers.

3. Make sure the result is real and convert to unsigned integer 8 (uint8 in MATLAB).

8.4.6 Compressing Modern Images

We have spent a lot of time on the Hungry Bee image. The reason for that is that it was close to being critically sampled, or at the Shannon sampling limit. You can notice that in Figure 8.17 we are able to achieve a compression rate of 16 without much loss. This is because the image is oversampled.

The good lesson from this is that with modern technology things are often oversampled. A simple cell phone picture is oftentimes excellent and samples way beyond the Shannon limit. There are ways to quantify this topic, but they are beyond the scope of the book. This falls into the space of information theory, and the Cramer–Rao bound gives us some guidance as to when there is no more information. The details of Cramer–Rao analysis are beyond the scope of this book, and the reader is encouraged to check these ideas out in either a basic statistics book, or an advanced image processing book.

Figure 8.16: The original image is shown at top left. We then show an image, which is subdivided and compressed with periodization at top right. The compression rate is 16. The image at the bottom right is that at the top right, but with edge effects suppressed somewhat. At bottom left, we have the image compressed and periodized, at rate of 4. The image is nearly identical and needed no additional processing.

Figure 8.17: The original, with a resolution of 2048 × 2048 at the top right. Top left, compression of 256. Bottom left, compression of 64. Finally, compression of 16. Note that the compressed image with a factor of 16 is still very good. This is because the image is oversampled.

8.5 Chapter Project:

1. Use convolution to find vertical and horizontal edges in a high-resolution picture such as the attached FishImage or Tiger.

2. Find directional edges (not just horizontal or vertical) and include a good example of these edges in your write-up.

3. Take a low-resolution image such as the attached Hungry Bee, and interpolate it so that it will be four times as large. Reorganize it as a .jpg image so that it can be displayed by MATLAB and saved. Include this in your write-up.

4. Try to downsample the image and then restore it to the higher resolution. This is a crude form of compression. Does this break down sooner or later? Include and analyze your results.

5. Take a 32×32 subsample of an image such as the attached Tiger, and investigate whether or not the 4-Flop technique from the book can reduce the number of significant coefficients needed at a 1.

Chapter 9
Medical Imaging

Medical imaging has revolutionized modern medicine in many ways. The first form of what we refer to as medical imaging is generally credited to Wilhelm Conrad Röntgen, who went on to win the first Nobel prize in 1901 for his efforts[19]. This first x-ray picture, which he took of his wife's hand clearly showing her hands and wedding ring is shown in Figure 9.1.

This first Nobel prize in physics was just a precursor of the many Nobel prizes that were to be attributed to medical imaging, or imaging in general. The Nobel institute lists no less than 14 Nobel prizes (out of 109 to date) in the imaging sciences [11]. Following Röntgen, Bloch and Purcell won the Nobel prize in 1952 for their contributions to Nuclear Magnetic Resonance, the predecessor of Magnetic Resonance Imaging today. In 1979, Alan Cormack and Godfrey Hounsfeld were awarded the Nobel Prize for their contributions to Computerized Tomography. In 2003, Lauterbur and Mansfield were awarded the Nobel prize for their contributions to Magnetic Resonance Imaging.

It is therefore hard to underestimate the impact that medical imaging and imaging in general has changed our understanding of the human body, and the universe in general in the last 120 years, which is very recent from a scientific perspective. We will not try to cover all of these developments, but will rather stick to two common modalities, and give a basic overview.

9.1 Computerized Tomography

X-ray imaging first demonstrated by Röntgen were the primary tool in medical imaging for nearly 70 years. They did not always reveal what one wanted to see, however, which was the 3-D structure inside the body. The 2-D pro-

© Springer Science+Business Media, LLC 2017
T. Olson, *Applied Fourier Analysis*,
https://doi.org/10.1007/978-1-4939-7393-4_9

jections or standard X-rays as in Figure 9.1 were valuable but primarily were used only for detecting fractures of bones. Very little use was cosidered for exploring the softer tissues of the body.Giving a shadow view of body.

Tomography is literally the study of 2-D slices of the human body. Medical imaging has become so common that we view this without much thought

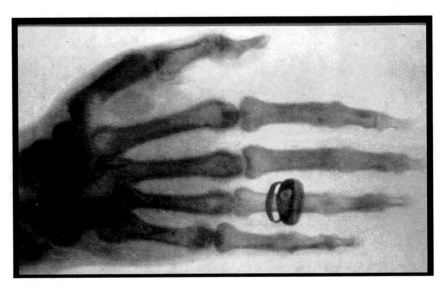

Figure 9.1: First x-ray picture of Mrs. Röntgen's hand, taken in 1896 by her husband Professor Röntgen. Röntgen went on to win the first Nobel prize in Physics for his work on x-rays.

as to the processes that make it possible. There are many different types of tomography, with different means for gathering the data. We will concentrate on the most basic of these applications. We assume throughout that we are imaging one slice of a three-dimensional density function $f(x, y, z)$. We will assume that z is fixed for each slice, so we will stick to two-dimensional densities, which you can think of as the density of tissues in one slice through your body.

9.1.1   Projections, or Line Integrals of a 2-D function

We begin by introducing notation for x-rays along a line in a particular plane, at a particular angle, and at a particular distance from the origin. We assume that the patient is in a confined area and that the origin in a two-dimensional x-y plane lies inside of the patient. This is illustrated in Figure 9.3. We now define the projections

$$P_f(r, \vec{\theta}) = \int f(r\vec{\theta} + t\vec{\theta}^{\perp})dt$$

which are the line integrals along an angle $\vec{\theta}^{\perp}$, at a distance r from the origin.

The main question is "Can we recover the function $f(x, y)$ from its projections $P_f(r, \vec{\theta})$?" The answer is yes. How we do this is another question. Proceeding somewhat historically, we consider the idea of backprojection. The question of associated with backprojection is: *When analyzing a two-dimensional function $f(x, y)$*, is the value at a single point perhaps associated with the average value of line integrals of the function through that point. Mathematically, we will define the line integrals through a point (x, y) in a nice form. To do this let us switch to vector notation. Let $\vec{x} = \langle x, y \rangle$. Note that if $\vec{x} = r\vec{\theta} + t\vec{\theta}^{\perp}$ then $\vec{x} \cdot \vec{\theta} = r$. Therefore, the line integral $P_f(\vec{x} \cdot \vec{\theta}, \vec{\theta}) \equiv P_f(r, \vec{\theta})$ is the line integral through $\vec{x}$ at angle $\vec{\theta}$.

The question then revolves around the equation, or approximation

$$f(\vec{x}) \approx \int P_x(\vec{x} \cdot \vec{\theta}, \vec{\theta})d\vec{\theta}? \tag{9.1}$$

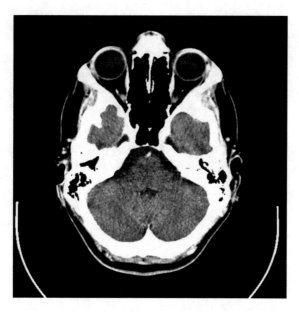

Figure 9.2:   We illustrate a computerized tomography picture of a slice through a human brain. Note that instead of the view from one side of the body, we are able to see the interior structure of the brain.

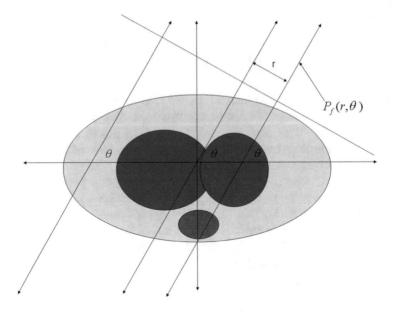

Figure 9.3: We illustrate the basics of the projection angles, coordinates, and vectors, r, θ, $\vec{\theta}$, and $\vec{x}$ above.

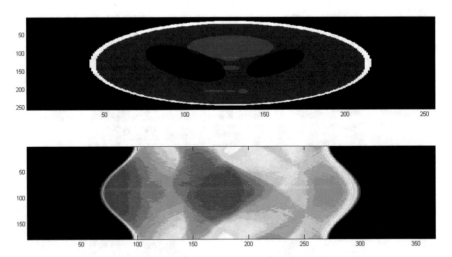

Figure 9.4: A standard test example, the Shepp–Logan phantom, is shown above. The projections which are often called sinograms are shown below.

Is this approximation valid with appropriate normalizing coefficients? Is it close? We first illustrate the backprojection process in Figure 9.5. We then show a counterexample in Figure 9.6.

Unfortunately, the backprojection process is not exact, or even a very good approximation. To understand why, consider Figure 9.6. All of the line integrals through the point $\vec{x}$ are non-zero and relatively large, so a simple average will not accurately detect the hole where the density is zero. To accurately reconstruct a two-dimensional density function $f(x,y)$, we must do some more mathematics, which turns out to be directly related to the Fourier Transform.

9.1.2 The Radon Tranform, or Central Slice Theorem

Backprojection is a nice intuitive idea, but unfortunately it is inaccurate, as our former example displayed. We now want to mathematically connect the one-dimensional projections

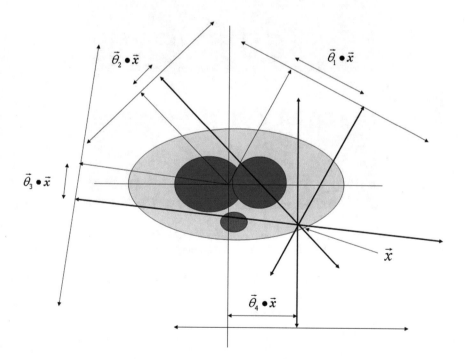

Figure 9.5: We illustrate the process of backprojection above. For each of the heavy lines, a line integral is obtained with a pencil thin x-ray, resulting in a measurement of the line integral. These line integrals are then added to give an approximation to the density at the point. In reality hundreds of measurements are used.

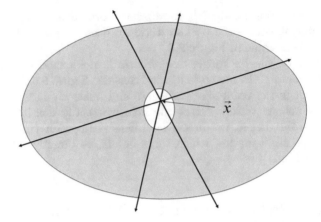

Figure 9.6: We illustrate the drawback of simple backprojection above. Although the density at the point $\vec{x}$ is zero, all line integrals through this point are non-zero, and rather large. A simple average will not accurately detect the hole.

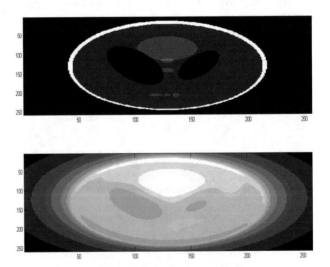

Figure 9.7: We illustrate the result of using simple backprojection on the Shepp–Logan phantom above. The result is similar to the original, but the small detail has been lost and there is much blurring.

$$P_f(\theta, r) = \int f(r\vec{\theta} + t\vec{\theta}^{\perp}) dt$$

to the two-dimensional function $f(x,y)$. To do this, we will move to the vector version of the two-dimensional Fourier Transform. The two-dimensional Fourier Transform is defined in rectangular coordinates by

$$\hat{f}(u,v) = \frac{1}{2\pi} \int \int f(x,y) e^{i(xu+yv)} dx dy.$$

But if we want to switch to vectors, we can define $\vec{u} = \langle u, v \rangle$ and $\vec{x} = \langle x, y \rangle$. We now have

$$\hat{f}(\vec{u}) = \frac{1}{2\pi} \int f(\vec{x}) e^{i\vec{x}\cdot\vec{u}} d\vec{x}, \tag{9.2}$$

which of course is identical to the above definition. The beauty of 9.2 is that we can input rather arbitrary vectors, as see what happens.

To this end, let us recall that the natural coordinates for tomography are $x = r\vec{\theta} + t\vec{\theta}^{\perp}$. Moreover $\vec{x}\cdot\vec{\theta}^{\perp} = r$. We want to see what the Fourier coefficients propagating at a fixed direction $\vec{\theta}$ have in common with the projections, so we consider

$$\begin{aligned}
\hat{f}(s\vec{\theta}) &= \frac{1}{2\pi} \int f(\vec{x}) e^{i\vec{x}\cdot s\vec{\theta}} d\vec{x} \\
&= \frac{1}{2\pi} \int \int f(r\vec{\theta} + t\vec{\theta}^{\perp}) e^{i(r\vec{\theta}+t\vec{\theta}^{\perp})\cdot s\vec{\theta}} dr dt \\
&= \frac{1}{2\pi} \int \int f(r\vec{\theta} + t\vec{\theta}^{\perp}) e^{i(rs)} dr dt \\
&= \frac{1}{2\pi} \int \int f(r\vec{\theta} + t\vec{\theta}^{\perp}) dt e^{i(rs)} dr \\
&= \frac{1}{2\pi} \int P_t(r, \vec{\theta}) e^{irs} dr. \tag{9.3}
\end{aligned}$$

Thus, a central slice of the two-dimensional Fourier Transform of $f(x,y)$, i.e., $\hat{f}(s\vec{\theta})$ can be obtained from the one-dimensional projections of the function or $P_f(r,\vec{\theta})$. Formally stated, we have

Theorem 9.1.1 (Radon Transform, or Central Slice Theorem) *The one-dimensional Fourier Transform of $P_f(\vec{\theta},r)$ is given by the central slice of the two-dimensional Fourier Transform, or $\hat{f}(s\vec{\theta})$. Mathematically,*

$$\mathcal{F}_1(P_f(r,\vec{\theta}) = \frac{1}{\sqrt{2\pi}} \hat{f}(s\vec{\theta}).$$

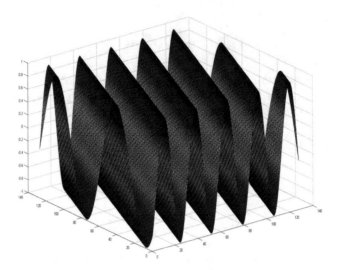

Figure 9.8: We illustrate one plane wave cosine function above. Note that it has a fixed propagation direction and frequency. Thus, if one takes a line integral of this extended function at a direction which is not perpendicular to its propagation direction, the result will be zero. If one takes a line integral exactly perpendicular to the cosine wave, the cosine wave will be revealed.

Understanding the Radon Transform

The Radon transform might seem mysterious to begin with, but it is very intuitive. In plain English, the Radon transform says that the projections of a function at a fixed angle $\vec{\theta}$ yield the Fourier components $\cos(s\vec{\theta})$ and $\sin(s\vec{\theta})$ which propagate in the direction $\vec{\theta}$. These are exactly functions which correspond to $\hat{f}(s\vec{\theta})$. Pictorially, consider one such function in Figure 9.8.

9.1.3 Filtered Backprojection

The Radon transform gives us a method to obtain the two-dimensional Fourier Transform of a density function $f(x, y)$ from one-dimensional line integrals. Once we have obtained the two-dimensional Fourier Transform, we can simply use the Inverse Fourier Transform, and we have the original density $f(x, y)$. Unfortunately this is not that easy. The Radon transform gives us measurements of the two-dimensional Fourier Transform, but on a grid which is very non-uniform. The Fourier Transform is sampled on a polar coordinate grid.

One approach is to try to interpolate this polar grid onto a standard rectangular grid and then use an inverse FFT on a standard rectangular grid to construct the density function. Unfortunately there are many difficult problems associated with this method. First, the interpolation is not easy, and second, the artifacts are very difficult to deal with.

A better approach is to realize that the polar coordinates are natural and therefore use polar coordinates. The Inverse Fourier Transform gives us

$$
\begin{aligned}
f(\vec{x}) &= \frac{1}{2\pi} \int \hat{f}(\vec{u}) e^{i\vec{x}\cdot\vec{u}} d\vec{u} \\
&= \frac{1}{2\pi} \int_0^\pi \int_{-\infty}^\infty \hat{f}(s\vec{\theta}) e^{i\vec{x}\cdot s\vec{\theta}} |s| ds d\theta.
\end{aligned} \tag{9.4}
$$

Now if we remember that $\vec{x} = r\vec{\theta} + t\vec{\theta}^\perp$, we see that

$$
\begin{aligned}
f(\vec{x}) &= \frac{1}{2\pi} \int_0^\pi \int_{-\infty}^\infty \hat{f}(s\vec{\theta}) e^{i\vec{x}\cdot s\vec{\theta}} |s| ds d\theta \\
&= \frac{1}{2\pi} \int_0^\pi \int_{-\infty}^\infty \hat{f}(s\vec{\theta}) |s| e^{irs} ds d\theta.
\end{aligned} \tag{9.5}
$$

The Inverse Fourier Transform of the product $\hat{f}(s\vec{\theta})|s|$ would be a convolution, but $|s|$ is not an element of $L^2(\mathbb{R})$. The reality is that we can put a finite window $w(s)$ which is wide enough to cover the bandwidth of $f(\vec{x})$, and have $|s|w(s) \in L^2(\mathbb{R})$. Thus, we have a convolution equation

$$
\begin{aligned}
f(\vec{x}) &\approx \frac{1}{2\pi} \int_0^\pi \int_{-\infty}^\infty \hat{f}(s\vec{\theta}) |s| w(s) e^{irs} ds d\theta \\
&= \int_0^\pi \left(P_f(r,\theta) * \mathcal{F}^{-1}(|s|w(s)) \right) (\vec{x}\cdot\vec{\theta}) d\theta.
\end{aligned} \tag{9.6}
$$

By choosing $w(s)$, appropriately we can make the error associated with the above approximation above arbitrarily small. If we denote $\mathcal{F}^{-1}(|s|w(s)) = k(r)$, then we have

$$
f(\vec{x}) = \int_0^\pi \left(P_f(r,\theta) * k(r) \right) (\vec{x}\cdot\vec{\theta}) d\theta. \tag{9.7}
$$

Thus, if we convolve or filter the projections with the appropriate convolution function, we can obtain an arbitrarily close approximation to the function $f(\vec{x})$ from its projections $P_f(\theta, r)$. Note that backprojection is used just as in Figure 9.5. The only addition to the process is convolution, or filtering, by the weighting function $k(r)$.

9.1.4   Non-locality of the Radon transform

We started out considering backprojection, and realized that we couldn't recover a function value at a point simply from the line integrals through that point. We had to use the filtering, or convolution function $k(r)$. The next logical question is "Can we recover the function from line integrals which nearly pass through that point?" The answer is unfortunately no.

To understand why, recall that $k(r)$ is the Inverse Fourier Transform of $w(s)|s|$, where $w(s)$ is some type of cutoff windowing function. The problem

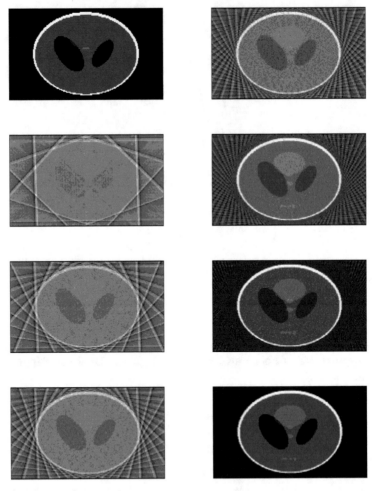

Figure 9.9:   We illustrate Filtered Backprojection with the images above. The image at the top left is the correct image. The image below it is the Filtered Backprojection image using only projections from 11 angles which are equally spaced. The progressively improving images have more and more angular projections, until we reach the last, which uses projections from 180 angles. Each projection has 367 individual line integrals.

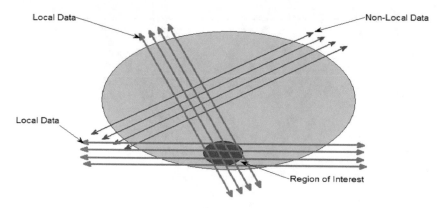

Figure 9.10: We illustrate the idea of the Region of Interest (ROI), local, and non-local data above. The Region of Interest (ROI), is in red, which might be an approximation for the spinal column of a patient. The green lines represent x-ray paths which pass through the ROI, and are therefore called local. The red lines represent x-ray paths which do not interest the ROI, and are therefore referred to as non-local. This project will minimize radiation along non-local x-ray paths.

with $|s|$ is that it is not differentiable at the origin. Since it is not differentiable at the origin, its Inverse Fourier Transform cannot decay quickly.

9.1.5 Localized Imaging

One problem with (9.7) is that the kernel $k(r)$ is very broad as a function of r, and as a result radiation measurements must be taken far from the Region of Interest. The reason for this kernel being broad is the jump discontinuity of the derivative of the function $|s|$ at the origin, from -1 to 1. Recall that $|s|$ is the necessary term due to the polar coordinates used in the Fourier inversion of the Filtered Backprojection formula (9.7). The basic theorems of Fourier Analysis dictate that this kernel cannot decay quickly.

We solve this problem by separating the discontinuity at the origin of $|s|$ into a separate portions $|s|w_2(s)$ at the origin, and $|s|(1-w_2(s))$ away from the origin. The corresponding Inverse Fourier Transforms will the a low-frequency kernel $k_l(r)$ which is the Inverse Fourier Transform of $|s|w_2(s)$, and a low-frequency kernel $k_h(r)$ which is the Inverse Fourier Transform of $|s|(1-w_2(s))$. This is illustrated in Figure 9.11. Thus, we have $k(r) = k_l(r) + k_h(r)$. The Filtered Backprojection algorithm now looks like

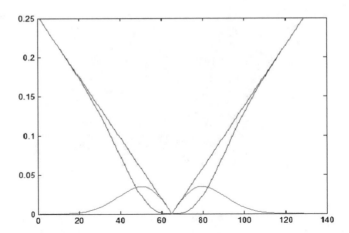

Figure 9.11: We show the correction term necessary for the Inverse Fourier Transform in polar coordinates above, i.e., $|s|$. The problem with $|s|$ is the jump discontinuity at the origin of the derivative of $|s|$, from 1 to negative 1. This dictates that the Inverse Fourier Transform of $|s|w(s)$, even with a suitably smooth window $w(s)$ will be very wide, rather than narrow. This can be solved as above, with one "low-frequency" term at the origin, and a "high-frequency" term. The corresponding decomposition of the kernel $k(r) = k_l(r) + k_h(r)$ will result in a low-frequency kernel as a function of radius, $k_l(r)$, which is not locally supported, and a high- frequency kernel $k_h(r)$ which is very narrow as a function of radius.

$$f(\vec{x}) \approx \int_0^\pi (P_f(r,\theta) * k(r))) \, (\vec{x} \cdot \vec{\theta}) d\theta, \tag{9.8}$$

$$= \int_0^\pi (P_f(r,\theta) * k_l(r)) \, (\vec{x} \cdot \vec{\theta}) d\theta + \int_0^\pi (P_f(r,\theta) * k_h(r))) \, (\vec{x} \cdot \vec{\theta}) d\theta, \tag{9.9}$$

$$= f_l(\vec{x}) + f_h(\vec{x}). \tag{9.10}$$

Thus, we will reconstruct the low- and high-frequency terms of $f(\vec{x})$ separately. The kernels are illustrated in Figure 9.12.

Initially, there seems to be no advantage to the change to two kernels $k_l(r)$ and $k_h(r)$ from a radiation reduction standpoint. The low-frequency kernel will require the gathering of large quantities of data from outside the region of interest. Thus, there is no apparent win in the fact that we can calculate the high-frequency component $f_h(\vec{x})$ from completely local measurements.

One must understand the structure of the projections, and corresponding structure of the Filtered Backprojection algorithms to see how to solve our problems with the low-frequency reconstruction $f_l(\vec{x})$. The structure theorem for the projections or Radon transform states that

$$P_f(\vec{\theta}, r) = (1 - r^2)^{-1/2} \sum_{l=0}^\infty T_l(r) h_l(\theta),$$

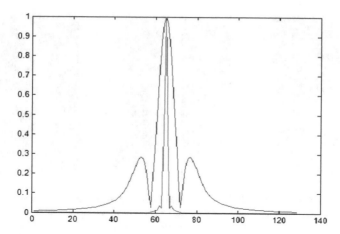

Figure 9.12: We show the kernel decompositions $k_l(r)$ and $k_h(r)$ above. The energy of $k_h(t)$ is contained within the interior 9 pixels of the current digitization or 9/512, to an accuracy of 1/10000. The energy concentration of $k_l(r)$, similarly measured, takes 175 terms. Thus, the low-frequency terms take a great deal of non-local information, and the high-frequency terms can be measured locally.

where $T_l(r)$ are the Chebyshev polynomials. Taking the Fourier Transform of this yields

$$\hat{f}(s\vec{\theta}) = \hat{P}_f(\theta, s) = \left(\frac{\pi}{2}\right)^{-1/2} \sum_{l=0}^{\infty} i^{-1} J_l(s) h_l(\theta) \tag{9.11}$$

$$= \left(\frac{\pi}{2}\right)^{-1/2} \sum_{l=0}^{N-1} i^{-1} J_l(s) h_l(\theta) + \left(\frac{\pi}{2}\right)^{-1/2} \sum_{l=N}^{\infty} i^{-1} J_l(s) h_l(\theta) \tag{9.12}$$

$$= \hat{f}_l(s, \theta) + \hat{f}_h(s, \theta), \tag{9.13}$$

where $J_l(s)$ are the Bessel functions, and $h_l(\theta)$ is an trigonometric polynomial of order l. The key to understanding (9.11) is that the low-frequency terms in s, which are the Bessel functions, are only multiplied in frequency by the low-order terms $h_l(\theta)$. Thus, we do not have to measure the low-frequency terms for many angles θ in order to accurately determine the complete low-frequency components of the image.

This was the approach of Localized Tomography espoused in [13, 16]. The sampling of the Projections, or Radon transform is illustrated in Figure 9.13. This approach is not easily implemented on standard fan-beam CT machines, however. The sampling is 0-1, meaning that the x-ray tube would either have to be shut off or modulated very quickly to accomplish this type of sampling. Secondly, the sampling is based on a Parallel beam geometry. It is even more difficult to imagine this 0-1 sampling in this case.

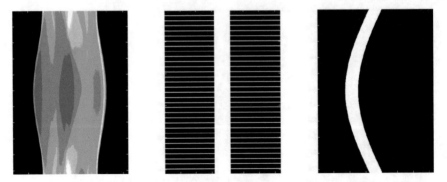

Figure 9.13:   At the left we illustrate a standard Radon transform or sinogram. We illustrate the sampling recommended for a central region of an image in [13, 16] in the center. While this greatly reduced the radiation levels by as much as 90% was designed for parallel beam geometries, and not completely feasible for fan-beam geometries. We illustrate the sampling which we are espousing for localized imaging in this project at the right. As opposed to the center sampling scheme, this is for an off-centered region. This sampling scheme will also easily work with fan-beam Geometries.

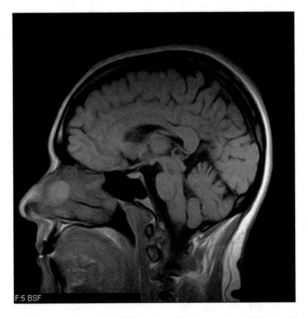

Figure 9.14:   We illustrate a typical MRI image of a human head above. Note the amount of detail of the soft tissues as well as bone in the image.[12].

9.2 Magnetic Resonance Imaging

Computerized tomography was a great advance in medical imaging, but like any device, it has its limitations. X-rays do a good job of penetrating the body, but they do present a radiation risk. Perhaps more importantly, they tend to penetrate soft tissue too readily, and as a result they are not sensitive small differences in the composition of soft tissues.

Magnetic Resonance Imaging (MRI) is a development which originated with Nuclear Magnetic Resonance (NMR). Felix Bloch and Edward Purcell received the Nobel prize for their contributions to NMR in 1952. NMR is fascinating and useful for a number of reasons. X-ray imaging relies on sending an x-ray through the body, and then measuring how much was absorbed by the body. NMR, on the other hand, is a process by which a material or the human body speaks to a very sensitive electronic receiver. The energy which is actually released by the body is specific to the varied atoms in the body.

Classic NMR was first used to study the chemical content of a material. The evolution of this process into a modern MRI machine for imaging the human body is certainly one of the great achievements of the twentieth century.

9.2.1 NMR and the Larmor Frequency: Making the Moments Sing

The Larmor frequency refers to the frequency at which the magnetic moment of a particular atom will spin in an external magnetic field B of a particular strength. While the physics of this phenomenon requires a great deal of understanding of electromagnetics and quantum mechanics, one can understand and work with magnetic resonance imaging from a simple, perhaps cartoon-level understanding.

Magnetic moments can be visualized as a center region (the atom), and a vector (the magnetic moment), such as in 9.15. If a material is not subjected to a large external magnetic field, the moments will be randomly arranged, as in Figure 9.15. The key to NMR or MRI is the use of a large electromagnetic field to align all of the moments in the direction of the field, as in Figure 9.16.

Figure 9.16 shows the moments in their equilibrium state. The key to NMR is to make these moments leave this equilibrium state temporarily. This is done by using a radio frequency signal which is tuned to the Larmor, or resonant, frequency of the atom. The atom then absorbs this energy, and the moment is pushed out of equilibrium much like a compressed spring. The atom will immediately try to return to equilibrium, and will do so by precessing in a helical manner such as in Figure 9.17.

The atom which is out of equilibrium is in a higher energy state than the atom in equilibrium, and as a result, the atom gives up this energy as it precesses back into equilibrium. The frequency of the energy which is given up is once again determined by the Larmor frequency. A key relationship is that the Larmor freqeuncy is directly related not only to the atom, but to the strength of the magnetic field. Thus, we can write

$$\omega = \lambda_a B,$$

where λ is associated with the atom, and B is the magnitude of the magnetic field.

The goal of standard NMR was to understand the chemical makeup of the material. Thus, the energy of different atoms was measured, and there was little concern about where in the sample material they were located. If a material was relatively uniform, the atoms would by definition be uniformly distributed within the sample. Magnetic Resonance Imaging, which was originally called Nuclear Magnetic Resonance Imaging, is an attempt to understand where different atoms are within a very non-uniform sample. The human body is the obvious application.

One obvious way to do this is to vary the magnetic field, so that the Larmor frequency will be different at each point in the body. Thus we would have

$$\omega(x,y,z) = \lambda_a B(x,y,z).$$

If the magnetic field is different at each point (x, y, z), then the amount of energy received at the corresponding frequency would be equivalent to the density of atoms at that location. Unfortunately this is not so easy. Magnetic fields are continuous, and as a result the magnetic field will be constant over

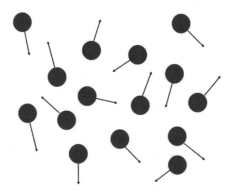

Figure 9.15: Magnetic moments of one atom in a typical material are shown above. They are not aligned and as a result are incoherent.

various level surfaces throughout the body. This is where things become more complicated, and much more interesting.

9.2.2   Relaxation times: T_1 and T_2 times

We will now go into more detail the data acquisition process and limitations. The first limitations which we must realize are the limitations which are emposed by T_1 and T_2 relaxation times. The T_1 relaxation time is the recovery to equilibrium, shown in Figure 9.17. This is the time that the magnetic moments take to return to being in line with the magnetic field after having been pushed out of equilibrium by the radio frequency pulse (RF pulse).

The T_2 time is the amount of time when a coherent signal can be gathered from the excited magnetic moments. The problem is that small anomalies in the field, which may be caused randomness in the actual material will make the individual magnetic moments precess at slightly above or below the main Larmor frequency. As a result, the magnetic moments no longer sing in harmony, or coherently. When the magnetic moments in precession have their coherent phases tend toward random phases, the signal that they are giving off drops off dramatically and becomes unusable.

The difference between T_1 and T_2 times is illustrated in Figure 9.18.

A fundamental limitation on an MRI system is illustrated in Figure 9.18. The first problem is that the signal can only be gathered before the signal becomes incoherent, or when T_2 has not decayed to zero. The second prob-

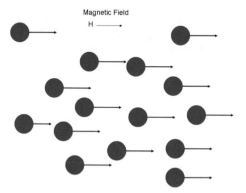

Figure 9.16:   When a strong magnetic field is applied to the material all of the magnetic moments of the atoms align. This is the equilibrium state for the material in the magnetic field.

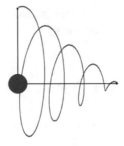

Figure 9.17:   Fundamental principle of Nuclear Magnetic Resonance (NMR), or Magnetic Resonance Imaging (MRI), is illustrated above. The magnetic moment is displaced from equilibrium by a radio frequency pulse. It then returns to equilibrium by spinning in a helical manner. It gives off the energy which is absorbed from the RF pulse along the way, at the appropriate Larmor frequency. The time that it takes to return to equilibrium is called the T_2 time.

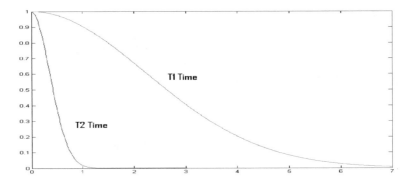

Figure 9.18: T_1 and T_2 times are shown above. The main problem is that the T_2 time is much shorter in general than the T_1 time. Therefore, one can only gather signal from the material or patient for a short time. In the mean time the material must be allowed to rest for the whole T_1 time to return to equilibrium. Thus, signal is only gathered for a short amount of the pulse excitation sequence.

lem is that all spins must return to equilibrium, which will put them back into phase, before a second RF pulse can be used to gather the next data acquisition. This is a fundamental problem, and much research has been done to try and speed up the time factors in MRI. Remember once again that by definition the patient is sick or injured and probably uncomfortable if an MRI is needed. Speeding this acquisition time is therefore very important.

Pulse Sequences

We now introduce what is known as the pulse sequences which are necessary to acquire data for MRI. A first most basic pulse sequence is illustrated in Figure 9.19. The basic idea is to give a timed sequence of events from the excitation to the acquisition which most efficiently gathers the signal.

9.2.3   MRI by way of Tomography

Imaging of the density at a single point in the body can be accomplished via tomographic methods. The standard method for doing this is to induce a magnetic field $B(x, y, z)$ which varies dramatically in one direction, which we will denote by z. Thus, at a specific point z_0, the field will be relatively constant, and therefore, we have reduced the problem to dealing with a two-dimensional slice of the body. Now in this slice, we will vary the magnetic field with respect to one of the variables, say x. Thus, in this slice, the magnetic field is constant with respect to y, or along vertical lines. Thus, when we inject energy at the Larmor frequency which is associated with a particular atom in this slice, all of the atoms along vertical lines will "sing" at the same frequency. By listening to this frequency, we will get a line integral along those lines of the density of that particular atom.

Now if we change the intensity of the magnetic field so that it is constant with respect to y, then we will get line integrals with respect to x. Tomographic imaging can be accomplished by altering the direction of these lines,

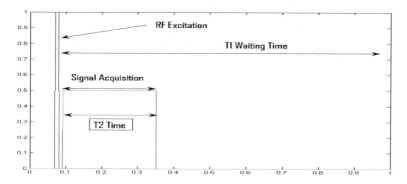

Figure 9.19: T_1 and T_2 times are shown above, in the context of a basic pulse acquisition time. The main problem is that the T_2 time is much shorter in general than the T_1 time. Therefore, one can only gather signal from the material or patient for a short time. In the mean time, the material must be allowed to rest for the whole T_1 time to return to equilibrium. Thus, signal is only gathered for a short amount of the pulse excitation sequence.

just as one does in tomography and the reconstruction techniques are much the same.

The problem with doing MRI using tomographic techniques is the same as it is doing traditional x-ray tomography. The sampling of the 2-D Fourier Transform is very non-uniform, with every set of line integrals passing through the origin. The result is that the highest frequencies, which will give highly sought after sharp edge detail, are the last to be properly sampled.

9.2.4 MRI via Phase Encoding

It turns out that MRI can be sampled much more efficiently by simply manipulating the magnetic gradients in order to directly sample read the 2-D Fourier Transform in a rectangular grid. Let us return to the 2-D Fourier Transform

$$\hat{f}(u,v) = \frac{1}{2\pi} \int \int f(x,y)e^{i(xu+yv)}\,dxdy,$$

where we assume that $f(x,y)$ is the density of a particular atom at the location (x,y). Remember that each atom will be "singing" to us at a particular Larmor frequency. The Fourier Transform is just the sum or integral of these weighted voices. Therefore, if we manipulate these voices so that they mimic the appropriate frequencies, we will be "hearing" the Fourier Transform of the image. If we have a uniform magnetic field through the slice which we have selected, then when we "bang" the magnetic moments out of equilibrium, they will "sing" to us and we will measure

$$\int \int f(x,y)e^{iw_L t}\,dxdy = e^{iw_L t} \int \int f(x,y)\,dxdy.$$

This give us the value of the Fourier Transform at $\hat{f}(0,0)$, which does tell us how much of the atom we are targeting is in the sample.

One question is "How do we get rid of the factor $e^{iw_L t}$ which is independent of x and y. There are two at least two possible answers. The first is a heterodyne, namely we mix the signal with an opposite phase signal $e^{-iw_L t}$, effectively cancel it. Thus, we truly have

$$\int \int f(x,y)\,dxdy.$$

The second possibility is to simply sample quickly so that t is essentially constant and $e^{iw_L t}$ has little effect. In all of these efforts, there are signal to noise tradeoffs which have been carefully treated in many places.

Pure Phase Encoding in x, and then in y

We want to sample the entire Fourier Transform $\hat{f}(u,v)$, however, not just $\hat{f}(0,0)$. Let us concentrate on sampling $f(u,v)$ for a fixed u_k. Thus, we want to sample

$$\hat{f}(u_k, v) = \frac{1}{2\pi} \int \int f(x,y) e^{i(xu_k + yv)} dx dy,$$

at an appropriate number of points v_k. For $v = 0$, we can accomplish this by varying the magnitude of the magnetic field B with the variable x. Note that the Larmor frequency is given by

$$\omega(x,y) = \lambda B(x,y).$$

Thus, if we vary the magnetic field $B(x,y)$ with linear gradient in the x direction, the Larmor frequency will be $\omega(x,y) = \lambda(B + cx)$. If we leave this gradient on for a finite amount of time δ_t, then the phase of this field will advance by $c\lambda x \delta_t$. We want this to be xu_k for any u_k so simple arithmetic allows us to choose $c\lambda \delta t = u_k$. After we turn off the gradient, the phase will be dependent upon x and y, and the Larmor frequency will once again be independent of x and y. Thus, we have spatially encoded the (x,y) plane with this phase, which is exactly what happens in the Fourier integral.

Thus, if we appropriately change the field with respect to x, for a finite amount of time, we induce the factor e^{ixu_k} in the sum and get the measurement

$$\int \int f(x,y) e^{ixu_k} e^{iw_L t} dx dy = e^{iw_L t} \int \int f(x,y) e^{ixu_k} dx dy.$$

Once again we can eliminate the term $e^{iw_L t}$ by a number of means.

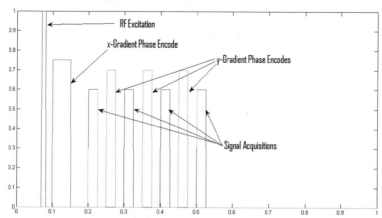

Figure 9.20: Above we illustrate the simple idea of acquiring data using pure phase encodes. The x-gradient is applied once to create the phase change in the x-direction. The y-gradient is turned on and off multiple times, creating multiple phase changes in y. Data is acquired multiple times, but the on–off process is limited by the T_2 time.

We do not only want the Fourier Transform for $v = 0$. We want to sample the Fourier Transform for fixed u_k and for different values of v_j, or

$$\hat{f}(u_k, v_j) = \frac{1}{2\pi} \int \int f(x, y) e^{i(xu_k + yv_j)} dx dy.$$

Let us assume that we have advanced the phase of the field in the x direction as above. Now we want to advance the phase of the field in the y direction. By the same arithmetic above, we turn on a magnetic gradient which causes the field to be $\omega(x, y) = \lambda(B + cy)$, and let $c\lambda\delta_t = v_j$. Thus, we can measure

$$e^{iw_L t} \frac{1}{2\pi} \int \int f(x, y) e^{i(xu_k + yv_j)} dx dy,$$

and use either a heterodyne or homodyne to eliminate the factor $e^{iw_L t}$.

Thus, by using two phase encodes, we can measure any one of the elements of the two-dimensional Fourier Transform. This can be repeated multiple times, as long as the T_2 time is not exceeded. This version is generally very slow. This is illustrated in Figure 9.20.

Phase Encoding in x and Frequency Encding in y

The problem with using pure phase encodeing steps is that you would need to do this individually for each coefficient, over and over again until all of the samples $\hat{f}(u_k, v_j)$ were obtained. One must remember that there is a sick patient in the machine, and the machine is expensive so time in the machine

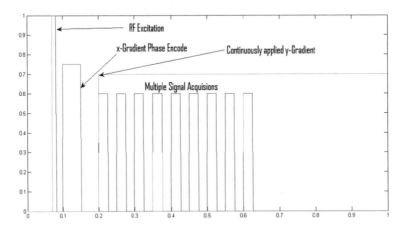

Figure 9.21: Above we illustrate the idea of acquiring data using phase and frequency encoding. We begin by creating an x-phase encode in the x-direction with a short x-gradient, which is then turned off. We then turn on the y-gradient, which effectively gives a continuous change in the y-phase. Thus, we can nearly continuously sample the y-phase changes, which corresponds to sampling a complete line through the two-dimensional Fourier Transform.

is expensive. A simple change to this basic methodology will speed this data acquisition up dramatically.

Instead of changing this one increment at a time, we will vary the gradient in the y direction continuously in time. This will allow us to sample a whole line of the Fourier Transform, $\hat{f}(u_k, v)$ in one step. This is accomplished by leaving the y-gradient on to induce the additional terms e^{iyv_k}. Leaving the y-gradient on allos, the phase to continue to advance. Thus, we sample

$$\int \int f(x,y)e^{i(xu_k+yv_j)}e^{iw_Lt}dxdy = e^{iw_Lt}\int \int f(x,y)e^{i(xu_k+yv_j)}dxdy,$$

for a number of values j in one excitation.

Note that we have just sampled one line of the 2-D Fourier Transform with one echo sequence. In general, resolutions of $512x512$ or higher are expected, so using 512^2 echo sequences is unacceptable, whereas 512 sequences is acceptable.

9.3   Chapter Project:

1. Utilize the MATLAB function phantom to create 256×256 phantom, perhaps the Shepp–Logan phantom.

2. Utilize the MATLAB function radon to create the projections, or sinogram, of the Shepp–Logan phantom created above. Can you see the sinusoidal tracts of the major density areas. If not perhaps examine the Shepp–Logan phantom in its modified state.

3. Utilize the MATLAB function iradon to create a reconstruction from the above sinograms. Try different numbers of angles, and different numbers of line integrals. How many do you think are necessary of each?

4. Use the MATLAB function as in Problem 3, and compare the choices of filters which are provided by MATLAB. Perhaps design one of your own.

5. **Limited Angle Tomography** Try only utilizing a limited angular range of the sinograms.... , i.e., start with only 45 degrees of projection data, then 90 degrees, and progress to 180 degrees. This should give you an idea of what the reconstruction algorithms are doing.

6. **Localized Tomography** Utilize only the central line integrals from the sinogram and compare against utilizing all of the sinogram, as in Figure 9.13. How large do you have to make the "kept" central portion of sinogram in order to get a good image.

Chapter 10
Partial Differential Equations

Fourier Analysis is central to the study of a great number of physical phenomena. The periodic nature of vibrations and wave propagation are generally modeled by differential equations. We will study just a few of the most basic differential equations here. There are many books which are exclusively devoted to this subject and the reader is encouraged to consider these after this basic introduction [2, 8, 21].

10.1 Ordinary Differential Equations: The Spring Equation

Ordinary differential equations are simply differential equations with only one variable. For instance, if an equation consists only of derivatives with respect to time, no matter how many derivatives are involved, the equation is referred to as an ordinary differential equation.

We begin with the most basic of differential equations, which is the spring mass system which is illustrated in Figure 10.1. The governing differential equation for this system is

$$y'' + ky = 0,$$

where y is the displacement of the mass from its natural rest position. The general solution to this system is given by

$$y(t) = c_1 \cos(\sqrt{k}t) + c_2 \sin(\sqrt{k}t).$$

The constants c_1 and c_2 are generally determined by the given initial conditions $y(0)$ and $y'(0)$. Substituting directly into the general solution yields

© Springer Science+Business Media, LLC 2017
T. Olson, *Applied Fourier Analysis*,
https://doi.org/10.1007/978-1-4939-7393-4_10

$$y(t) = y(0)\cos(\sqrt{k}t) + \frac{y'(0)}{\sqrt{k}}\sin(\sqrt{k}t).$$

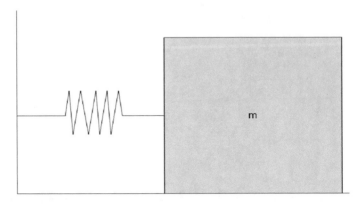

Figure 10.1: We illustrate a simple spring mass system above. The mass is situated on a nearly frictionless surface, such as ice. The goal is to model the behavior of the mass when it is displaced from its natural rest position.

While this is a very basic ordinary differential equation, it will be at the center of the development of the more general partial differential equations which we are now going to study.

10.2   The Wave Equation

We will now begin our discussion of the partial differential equations with two simple examples. The first is the bound string equation, which models the behavior of a string under tension, as is the case with a guitar or violin string.

10.2.1   The Bound String Equation

The question is "How does the string move after it has been displaced, perhaps by plucking it?". To begin with, we must describe the location of the string at any given point in time. This is generally done by an equation $u(t, x)$, which is the displacement from rest of the spring at time t, and location x. Moreover, we assume that $u(t, 0) = u(t, L) = 0$ where L is the length of the string since the string is bound at both endpoints. Finally, we assume that we know the location and velocity of the string at time 0, or we are given $u(0, x) = f(x)$, and $u_t(0, x) = g(x)$, where $u_t(t, x) = \frac{\partial}{\partial t} u(t, x)$.

The basic partial equation which governs the motion of this string is given

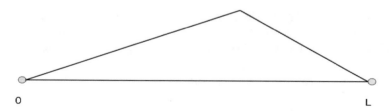

O   L

Figure 10.2:   We illustrate a simple bound string problem above. The string is flexible and under tension.

by

$$\frac{\partial^2}{\partial t}\left(u(t, x)\right) \equiv u_{tt} = c^2 u_{xx} \equiv \frac{\partial^2}{\partial x}\left(u(x, t)\right).$$

Collecting the governing, boundary, and initial equations gives us the system

$$u_{tt} = c^2 u_{xx} \qquad\qquad (10.1)$$
$$u(0, x) = f(x)$$
$$u_t(0, x) = g(x)$$
$$u(t, 0) = u(t, L) = 0.$$

Note that the constant c is determined by a variety of parameters, including (a) the type and thickness of string, (b) the tension on the string. When we tune a violin or guitar, we increase or decrease the tension on the string, which changes c.

10.2.2   Separation of Variables

The standard way to solve this equation is to assume that solution of this equation can be written as the product of two functions of one variable, or that individual solutions can be found which are of the form $u(t, x) = T(t)X(x)$. While this is just an assumption, we will find out that this method will indeed produce enough solutions to solve the system 10.1. Plugging our assumption into the system yield the equations

$$T''(t)X(x) = c^2 T(t)X''(x)$$

which can be rewritten as

$$\frac{T''(t)}{c^2 T(t)} = \frac{X''(x)}{X(x)}. \tag{10.2}$$

Note that the left side of the equation depends only upon t, and the right side of the equation depends only upon x. Thus, changing t on the left will not change the right side, and as a result, both sides much be constant. Thus, we have

$$\frac{T''(t)}{c^2 T(t)} = \frac{X''(x)}{X(x)} = \eta. \tag{10.3}$$

Now we can deal with each of these separately, yielding the two equations

$$T''(t) - c^2 \eta T(t) = 0$$
$$X''(x) - \eta X(x) = 0,$$

where k is an arbitrary constant. We now want to consider different values for η. If $\eta > 0$ the solutions to the equations would be of the form $\exp(\sqrt{\eta}t)$, which does not make physical sense, because the system would blow up as $t \to \infty$. Therefore, we assume that $\eta < 0$, or we simply choose to consider the equations

$$T''(t) + c^2 \eta T(t) = 0$$
$$X''(x) + \eta X(x) = 0,$$

which have general solutions

$$T(t) = a_1 \cos(c\sqrt{\eta}t) + a_2 \sin(c\sqrt{\eta}t)$$
$$X(x) = b_1 \cos(\sqrt{\eta}x) + b_2 \sin(\sqrt{\eta}x).$$

A simple check shows that, indeed, the function $u(x,t) = T(t)X(x)$ is a solution of the governing equation for any arbitrary value of η.

At this point in time, we are going to assume for simplicity that $L = \pi$. Now note that since $u(0,t) = u(\pi,0)$ it seems natural that we might only need the sin terms for $X(x)$. We need to be able to meet the initial conditions $u(0,x) = f(x)$ and $u_t(0,x) = g(x)$ and this seems unlikely for any one function $X(x)T(t)$ above. We have that $\eta > 0$ is arbitrary, however, so we would like to choose a subset of solutions such that we can meet the initial conditions. If we choose $\eta = k^2$, then we have the functions $u_k(t,x) = \sin(kx)(a_1 \cos(ckt) + a_2 \sin(ckt))$. The question now is whether these are enough, or is the general solution

$$u(t,x) = \sum_{k=1}^{\infty} u_k(t,x) = \sum_{k=1}^{\infty} \sin(kx)\left(a_1(k)\cos(ckt) + a_2(k)\sin(ckt)\right) \tag{10.4}$$

sufficient to satisfy the initial conditions?

Plugging directly into this sum we get

$$u(0, x) = \sum_{k=1}^{\infty} \sin(kx)\, (a_1(k) \cos(0) + a_2 \sin(0)) = \sum_{k=1}^{\infty} a_1(k) \sin(kx) = f(x),$$

and

$$u_t(0, x) = \sum_{k=1}^{\infty} \sin(kx)\, (a_1(k)(-ck) \sin(0) + a_2(ck) \cos(0)) = \sum_{k=1}^{\infty} (ck) a_2(k) \sin(kx) = g(x).$$

$$(10.5)$$

Recall that these are sine expansions on the half interval $[0, \pi]$ which implies that we can indeed satisfy the initial conditions, by utilizing the appropriate sine expansions.

Returning to the formulas for the coefficients of a sine series, we have that

$$a_1(k) = \frac{2}{\pi} \int_0^{\pi} f(x) \sin(kx) dx,$$

and

$$ck a_2(k) = \frac{2}{\pi} \int_0^{\pi} g(x) \sin(kx) dx,$$

or

$$a_2(k) = \frac{2}{ck\pi} \int_0^{\pi} g(x) \sin(kx) dx.$$

10.2.3   Revisiting Completeness and Sine Transforms

The above analysis can sometimes be taken for granted. We would like to emphasize that it highlights two of the most important properties of Fourier Series. First, that sines and cosines occur very naturally in science, nature, and engineering. While we have chosen a very basic example, sines and cosines end up giving approximate solutions to nearly all vibrational problems. These occur in engineering, physics, and many other sciences.

The second, which sometimes overlooked, is that the completeness of the sine transform allows us to write the general solution to this basic PDE in the form (10.4). Recall that the basic functions $1, \cos(kx), \sin(kx)$ form a complete set of orthogonal functions on $[-\pi, \pi]$. Furthermore recall that on $[0, \pi]$, we can choose to use either a cosine expansion, by artificially making the function even, or a sine expansion, by artificially making the series odd. Since the string has the boundary conditions $u(t, 0) = u(t, \pi) = 0$, it is more natural to choose the sine expansion since the sine functions are zero at 0 and

π. It is worth noting that we could use a cosine expansion. The convergence of this sequence would be very slow since every cosine would be 1 at the boundaries.

The number of choices afforded to this simple problem highlights the power of Fourier Series and Fourier Analysis.

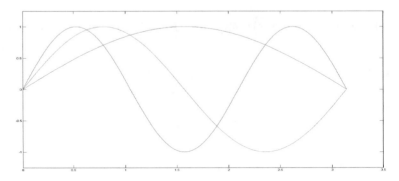

Figure 10.3: We illustrate the first eigenfunction of a basic string and the second and third eigenfunctions. Musically, these are sometimes referred to as overtones. Note that in the time that the first eigenfunction completes one vibrational cycle, the second and third eigenfunctions will complete two and three cycles, respectively.

10.2.4   Eigenfunctions: Musical Overtones

An outcome of the general solution to the wave equation 10.4 is that we have the eigenfunctions of the string. To understand why we call them eigenfunctions consider the individual functions

$$u_k(t, x) = \sin(kx)\cos(ckt),$$

which are involved in the expansion when the initial displacement $f(x) \neq 0$, and $g(x) = 0$. Note that as t evolves, the shape of $\sin(kx)$ is not changed, but rather $\cos(ckt)$ modulates $\sin(kx)$ for each individual t. Just as an eigenvector does not change anything except its length or direction, $\sin(kx)$ is an eigenfunction which does not change its shape. The function $\cos(ckt)$ is a continually changing eigenvalue, for each individual t.

This explains, to some extent, the basics of the wave equation. Simple waves are not changed in shape. Their magnitude changes. In addition, the speed with which it changes is related via

$$u_{xx} = c^2 u_{tt}$$

making high order sinusoids change much faster than the low-order sinusoids.

Musically, the respective sinusoids or eigenfunctions are referred to as over-tones. Each overtone increases in frequency over the last. We give you a few of these examples in Figure 10.3.

Solving the String Equation numerically

The first key to solving the string equation is obviously the computation of the sine coefficients of $f(x)$ and the modified coefficients of $g(x)$. We discussed the necessary steps for this in Chapter 3. One must first extend $f(x)$ and $g(x)$ to be odd functions. The Fast Fourier Transform should then only contain sine coefficients, or be purely imaginary. Once those coefficients have been computed, they can be stored for use at every time period.

The value of the evolving wave $u(t, x)$ at any time is then computed by multiplying these sine coefficients by their respective weights $\cos(ckt)$, for each frequency k. Making sure these are appropriately matched is the only difficult detail and should be checked with the simple sinusoids first. An inverse Discrete, or Fast Fourier Transform will then compute the values of the $u(t, x)$ for each time period. Displaying them as a time-evolving sequence gives one a feeling for the evolution of the wave.

Two examples:

We illustrate two examples of the string equation in Figure 10.4. On the left, we have utilized a hat function, which models a string plucked in the middle. Note that the string evolves in a very strict, nearly linear manner. This can be explained by examining the governing equation $u_{tt} = c^2 u_{xx}$. When the equation $u(t, x)$ is linear in a region, $u_{xx} = 0$, and therefore, $u_{tt} = 0$. With no starting velocity, the string remains constant in these regions until some-thing changes this condition. This happens starting from the center where the second derivative is either large (or technically might not exists) and moves outward.

The second example comes from the initial function $f(x) = \sin(x) + 2\sin(2x) - 3 * \sin(3x)$. This evolves much more fluidly, in a manner which we might think is typical of a wave, since these are overtones of the string.

10.2.5   Problems and Exercises:

1. Create a small program which will display the evolution of the basic waveforms with time, using MATLAB or a similar program. The wave should smoothly evolve in time, giving one a feeling for its true time evolution.

2. Similarly to Problem 1, let the initial function $f(x)$ be a simple linear combination of several overtones, and look at the evolution of the function. You may have to adjust the limits of the graph.

3. [**Project**] Write a program that will accept an initial sampled vector approximating $f(x)$, and assume that $g(x) = 0$. Allow the using to input the time steps and x resolution. The program should produce a time-evolving graph of the wave, as described above.

4. [**Project**] Write an extended version of the program in Exercise 3 which will allow the user to input a discrete, sampled $f(x)$, and a discrete, sampled $g(x)$. The program should produce a time-evolving graph of the wave, which may be more complicated than in 3.

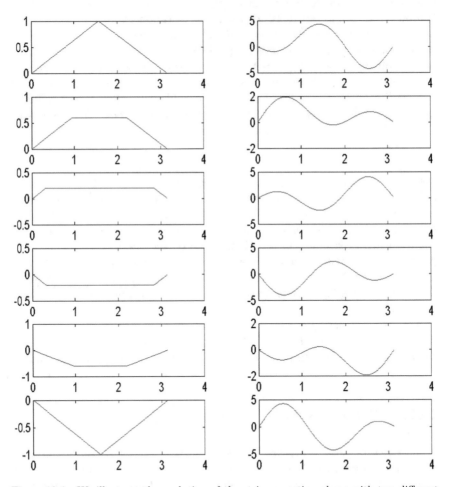

Figure 10.4: We illustrate the evolution of the string equation above with two different initial values $f(x)$. Two different initial starting conditions are shown at the top. On the left is a realistic model of a string which has been plucked in the middle. Note that its evolution is primarily in a rather rigid, semi-linear form. On the right, the string evolves in a very fluid manner. The initial starting conditions at the top on the right consists of a couple of basic sinusoids, or overtones.

5. **[Project]** With the programs from either Problems 3 or 5, use an input
   $f(x)$ which is similar to that in Figure 10.2. Does it seem to match
   what you see in the governing differential equation $u_{tt} = c^2 u_{xx}$?

6. **[Project]** With the programs from either Problems 3 or 5, use an input
   $f(x)$ which is somewhat arbitrary and may not be physically realistic. Does it seem to match what you see in the governing differential
   equation $u_{tt} = c^2 u_{xx}$? Does this tell you anything?

10.2.6   Two-Dimensional Wave Equation: The Square Drum

The wave equation on a square drum is a simple extension of the above
analysis. We have the drum, which we will assume for simplicity to be on a
coordinate system so that the boundary is given by $[0, \pi]x[0, \pi]$. We let the
displacement of the drum from rest at time t be $u(t, x, y)$, and assume that the
rest condition is $u(t, x, y) = 0$. Furthermore, we assume that the boundaries
are given by $u(t, 0, y) = u(t, x, 0) = u(t, \pi, y) = u(t, x, \pi) = 0$. We assume that
the initial position is given by $u(0, x, y) = f(x, y)$, and the initial velocity
$u_t(0, x, y) = g(x, y)$. The governing equation for the drum is the analogue of
the string

$$u_{tt} = c^2 \left(u_{xx} + u_{yy} \right).$$

Thus, we can sum up the system by the equations

$$u_{tt} = c^2 \left(u_{xx} + u_{yy} \right) \tag{10.6}$$
$$u(t, 0, y) = u(t, x, 0) = u(t, \pi, y) = u(t, x, \pi) = 0$$
$$u(0, x, y) = f(x, y)$$
$$u_t(0, x, y) = g(x, y)$$

We proceed to look for individual solutions by considering separated solutions of the form $u(t, x, y) = T(t)X(x)Y(y)$. Substituting into the governing
equation 10.6 and dividing by $T(t)X(x)Y(y)$, we get

$$\frac{T''(t)}{c^2 T(t)} = \frac{X''(x)}{X(x)} + \frac{Y''(y)}{Y(y)}.$$

All three of the fractions in the above equation must be constant, so we have

$$\frac{X''(x)}{X(x)} = -\mu,$$

$$\frac{Y''(y)}{Y(y)} = -\eta,$$

$$\frac{T''(t)}{c^2 T(t)} = -(\mu + \eta),$$

where we have assumed that μ and η are positive, to avoid exponentially growing solutions as in the string equation.

Thus we have the equations

$$X''(x) + \mu X(x) = 0,$$
$$Y''(y) + \eta Y(y) = 0,$$
$$T''(t) + c^2(\mu + \eta)T(t) = 0.$$

The corresponding basic solutions are therefore

$$u_{\mu,\eta}(t, x, y) = \sin(\sqrt{\mu}x) \sin(\sqrt{\eta}y)(a \cos(c(\sqrt{\mu + \eta})t) + b \sin(c(\sqrt{\mu + \eta})t).$$

We have used only sine terms for the $X(x)$ and $Y(y)$ components because of the boundary conditions, similarly to the string equation expansion. Direct substitution into 10.6 will verify that these are indeed solutions for every μ and η.

Now we want to once again choose a rich enough collection of functions, and this is accomplished by letting $\mu = m^2$, and $\eta = n^2$, giving the solutions

$$u_{m,n}(t, x, y) = \sin(mx) \sin(ny) \left(a_{m,n} \cos(c(\sqrt{m^2 + n^2})t) + b_{m,n} \sin(c(\sqrt{m^2 + n^2})t)\right).$$

To simplify notation we let $\zeta = \sqrt{m^2 + n^2}$ to get

$$u_{m,n}(t, x, y) = \sin(mx) \sin(ny) \left(a_{m,n} \cos(\zeta t) + b_{m,n} \sin(c\zeta t)\right).$$

We would like our final solutions to be of the type

$$u(t, x, y) = \sum_{m=0}^{\infty} \sum_{n=0}^{\infty} u_{m,n}(t, x, y)$$

$$= \sum_{m=0}^{\infty} \sum_{n=0}^{\infty} \sin(mx) \sin(ny) \left(a_{m,n} \cos(\zeta t) + b_{m,n} \sin(c\zeta t)\right). \qquad (10.7)$$

The boundary conditions are automatically met by the choice of the sine functions in x and y. The initial conditions imply that

$$f(x, y) = u(0, x, y) = \sum_{m=0}^{\infty} \sum_{n=0}^{\infty} a_{m,n} \sin(mx) \sin(ny),$$

and

$$g(x,y) = u_t(0,x,y) = \sum_{m=0}^{\infty} \sum_{n=0}^{\infty} c\zeta b_{m,n} \sin(mx) \sin(ny).$$

Thus, we need two-dimensional sine expansions of $f(x,y)$ and $g(x,y)$ to find the coefficients $a_{m,n}$, and $b_{m,n}$.

These can either be calculated directly as

$$a_{m,n} = \frac{4}{pi^2} \int_0^{\pi} \int_0^{\pi} f(x,y) \sin(mx) \sin(ny) dx dy,$$

and

$$b_{m,n} = \frac{4c\zeta}{pi^2} \int_0^{\pi} \int_0^{\pi} f(x,y) \sin(mx) \sin(ny) dx dy.$$

Computing the 2-D sine series with the 2-D FFT

In the discrete case, you can compute these sine coefficients much as we have computed sine a cosine series discretely before. Note that anything of the form $p(x,y) = \sin(mx) \sin(ny)$ has the properties $p(x,y) = -p(-x,y)$, $p(x,y) = -p(x,-y)$, and $p(x,y) = p(-x,-y)$. Thus, this gives a method for taking a grid on $[0,\pi] \times [0,\pi]$, and completing it to $[-\pi,\pi] \times [-\pi,\pi]$. From this grid, which is 4 times the size of the original grid, we can use the standard two-dimensional Fourier Transform to calculate the sine series. Multiplying by the appropriate weightscos(ζt) and $\sin(\zeta t)$ will yield the desired result.

10.2.7   Problems, Exercises, and Projects

1. Create a small program which will display the evolution of the basic waveforms with time, using MATLAB or a similar program. The wave should smoothly evolve in time, giving one a feeling for its true time evolution.

2. Similarly to Problem 1, let the initial function $f(x)$ be a simple linear combination of several overtones, and look at the evolution of the function. You may have to adjust the limits of the graph.

3. **[Project]** Write a program that will accept an initial sampled vector approximating $f(x)$, and assume that $g(x) = 0$. Allow the user to input the time steps, and t-x resolution. The program should produce a time-evolving graph of the wave, as described above.

4. **[Project]** Write a program that will accept an initial sampled vector approximating $f(x)$, with an arbitrary $g(x)$. Allow the user to input

the time steps, and t-x resolution. The program should produce a time-evolving graph of the wave, as described above.

5. **[Project:2-D]** Write an extended version of the program in Exercise 3 which will allow the user to input a discrete, sampled $f(x, y)$, and a discrete, sampled $g(x, y)$. The program should produce a time-evolving graph of the wave, which may be more complicated than in 3.

6. **[Project]** With the programs from either Problems 3 or 5, use an input $f(x)$ which is similar to that in Figure 10.2. Does it seem to match what you see in the governing differential equation $u_{tt} = c^2 (u_{xx} + u_{yy})$?

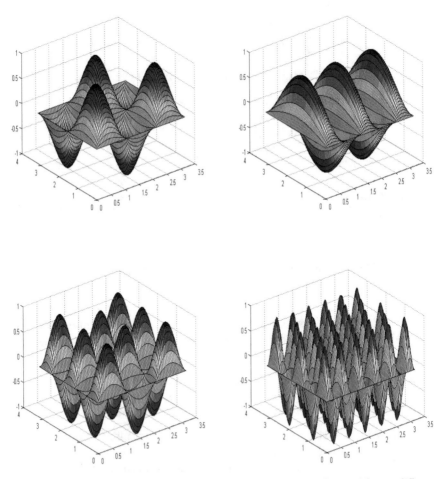

Figure 10.5: We illustrate the evolution of the string equation above with two different initial values $f(x)$. Two two initial starting points are shown at the top. On the left is a realistic model of a string which has been plucked in the middle. Note that its evolution is primarily in a rather rigid, semi-linear form. On the right, the string evolves in a very fluid manner. The starting point consists of a couple of basic sinusoids, or overtones.

7. **[Project]** With the programs from either Problems 3 or 5, use an input $f(x)$ which is somewhat arbitrary, and may not be physically realistic. Does it seem to match what you see in the governing differential equation $u_{tt} = c^2(u_{xx} + u_{yy})$? Does this tell you anything?

10.3   The Heat Equation

10.3.1   A One-Dimensional Conductor

We will now study the simple one-dimensional heat equation. This models the conduction of heat in a long thin rod. We will assume that the rod is kept at a steady temperature at the endpoints. We will assume that this influence is the only external force on the system. For a more complicated version, one might think of having cooling tubes through an engine, which attempt to control the temperature of the engine. This is of course much more difficult, and very valuable information but is left for another course.

We assume that the rod has finite length L, and let the heat at time t and position x along the rod be given by $u(t, x)$. The governing differential equation for the system is given by $u_t = cu_{xx}$, where c is a constant which represents the conductivity of the material in the rod. Recall that materials have different conductivities, and are used appropriately. Some materials insulate, which would imply that c is small, resulting in little change in the heat. Some conduct heat very well, such as metals, and as a result, c is large.

We assume as stated above that $u(t, 0) = T_1$, and $u(t, L) = T_2$, where T_1 and T_2 are two fixed temperatures. One might think of a rod which has one end in ice water, at 0 degrees Celsius, and the other in boiling water, at 100 degrees Celsius. We also assume that the initial temperature $u(0, x) = f(x)$ is given. Thus the system is determined by

$$u_t = cu_{xx} \qquad (10.8)$$
$$u(t, 0) = T_1$$
$$u(t, L) = T_2$$
$$u(0, x) = f(x).$$

From the beginning, we can observe that there is a simple solution to this system, if we ignore the initial condition $f(x)$. Let us consider a line between T_1 and T_2, namely the function

$$s(x) = \frac{T_2 - T_1}{L}x + T_1, \qquad (10.9)$$

we see that it satisfies the governing equation. Quite simply if $u(t,x) = s(x)$, then $u_{xx} = 0$ for all t, and $u_t = 0$ which satisfies the equation. We also have that $u(t,0) = T_1$, and $u(t,L) = T_2$ so the boundary conditions are satisfied. This solution does not change with time. Therefore, we call this the steady state solution. We will see later that this is always the long term, or steady state solution. Without any other external stimulation the system will always tend to this solution, or $\lim_{t \to \infty} u(t,x) = s(x)$. This solution will not satisfy the initial condition $f(x)$, unless of course $f(x) = s(x)$.

We now want to consider an alternate system to 10.8. We will consider

$$u_t = cu_{xx} \tag{10.10}$$
$$u(t,0) = 0$$
$$u(t,L) = 0$$
$$u(0,x) = f(x) - s(x) = h(x).$$

where $s(x)$ is defined by 10.9.

We proceed as with the solution to the string problem and assume that we can find sufficient solutions of the type $u(t,x) = T(t)X(x)$. Substituting in to the governing equation, we get

$$T'(t)X(x) = cT(t)X''(x)$$

and once again have a separated equations

$$\frac{T'(t)}{cT(t)} = \frac{X''(x)}{X(x)} = \eta,$$

where η is an arbitrary unknown constant.

This produces the two equations

$$T'(t) - c\eta T(t) = 0 \tag{10.11}$$
$$X''(x) - \eta X(x) = 0. \tag{10.12}$$

Now, as in the string equation, we will consider the possible values for η and choose using physical reality the possible solutions. The solutions for $T(t)$ are quite simply $T(t) = ae^{c\eta t}$, where a is an arbitrary constant. If η where positive, then this system would tend toward infinity and thus we assume that $\eta \le 0$. If $\eta = 0$, then we have a constant temperature, which is a trivial solution so we assume that $\eta < 0$. Reflecting on this, let us change these equations to

$$T'(t) + c\eta_1 T(t) = 0 \tag{10.13}$$
$$X''(x) + \eta_1 X(x) = 0. \tag{10.14}$$

where we assume that $\eta_1 > 0$. Our individual solutions will now be of the form

$$u(t, x) = T(t)X(x) = e^{-c\eta_1 t} \left(a_1 \sin(\sqrt{\eta_1}t) + a_2 \cos(\sqrt{\eta_1}t))\right).$$

Once again, we want a set of solutions which is rich enough to satisfy the initial conditions, with a series expansion. We can therefore let $\sqrt{\eta} = k\pi/L$, and we have the solutions

$$u_k(t, x) = T(t)X(x) = e^{-c\eta_1 t} \left(a_1 \sin\left(\frac{k\pi t}{L}\right) + a_2 \cos\left(\frac{k\pi t}{L}\right)\right).$$

If we assume that the rod obeys the initial and boundary conditions, then we have $f(0) = T_1$, and $f(L) = T_2$, or $h(0) = h(L) = 0$. Utilizing the Sine expansion for $h(x)$, then suffices, and we get

$$u_{init}(t, x) = \sum_{k=1}^{\infty} a(k)e^{-c\eta_1 t} \sin\left(\frac{k\pi t}{L}\right),$$

where

$$a_k = \frac{2}{L} \int_0^L h(x) \sin\left(\frac{k\pi t}{L}\right).$$

The solution $u_{init}(t, x)$ is sometimes called the transient solution, since it dies away at $t \to \infty$.

Now our final solution to 10.8 is given by

$$u(t, x) = s(x) + u_{init}(t, x),$$

where $s(x)$ is the steady state solution which satisfies the boundary conditions, and $u_{init}(t, x)$ is the transient solution which satisfies the initial condition.

We note at this time a fundamental difference between the heat equation and the one-dimensional string equation. The solution to the heat equation moves rather quickly to the steady state solution, while the string equation continues to evolve without long-term damping. Some of this is due to the nature of the equations. The heat equation will move to the steady state solution in reality. The string will eventually move back to the steady state solution $u(t, x) = 0$. The reason why the solution which we obtained does not do this is because damping is not included in the equations.

10.3.2   Eigenfunctions and Examples:

First, we start with an example where the endpoints are both held at 0, and the initial function is $f(x) = \sin(x) + 2\sin(2x) - 3 * \sin(3x)$ on $[0, pi]$. The second example has $f_1(x) = f(x) + 5x/\pi$. Note that in this case, $s(x) = 5x/\pi$ is the steady state solution. Snapshots of the evolution of these solutions are shown in Figure 10.6.

Note that in these cases, the eigenfunctions are once again $\sin(kx)$, but that they simply evolve to zero with the eigenvalues e^{-ckt}.

10.3.3   A Two-Dimensional Example: The Heat Equation on a Square Plate

The heat equation on a square plate is an extension of the above analysis. We have a metal plate, which we will assume for simplicity to be on a coordinate system so that the boundary is given by $[0, \pi] x [0, \pi]$. We let the temperature at time t be $u(t, x, y)$. Furthermore, we assume that the boundaries are given by $u(t, 0, y) = u(t, x, 0) = u(t, \pi, y) = u(t, x, \pi) = 0$. We assume that the initial temperature is given by $u(0, x, y) = f(x, y)$. The governing equation for the plate is the analogue of the long wire

$$u_t = c^2 \left(u_{xx} + u_{yy} \right).$$

Thus, we can sum up the system by the equations

$$u_t = c^2 \left(u_{xx} + u_{yy} \right) \tag{10.15}$$
$$u(t, 0, y) = u(t, x, 0) = u(t, \pi, y) = u(t, x, \pi) = 0$$
$$u(0, x, y) = f(x, y)$$

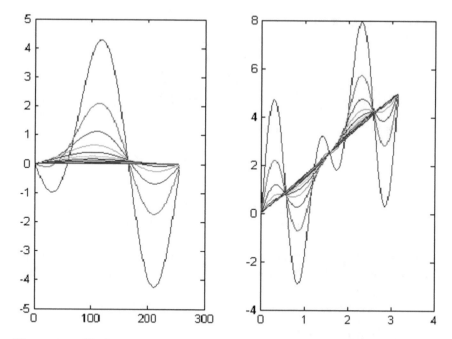

Figure 10.6: We illustrate the evolution of the heat equation above. On the left, both endpoints are kept at 0, so the evolution is to $u(x,t) = 0$. On the right, the endpoints are kept at 0 and 5, so the evolution is to the steady state $s(x) = 5x/\pi$.

We proceed to look for individual solutions by considering separated solutions of the form $u(t,x,y) = T(t)X(x)Y(y)$. Substituting into the governing equation 10.15 and dividing by $T(t)X(x)Y(y)$, we get

$$\frac{T'(t)}{c^2 T(t)} = \frac{X''(x)}{X(x)} + \frac{Y''(y)}{Y(y)}.$$

All three of the fractions in the above equation must be constant, so we have

$$\frac{X''(x)}{X(x)} = -\mu,$$

$$\frac{Y''(y)}{Y(y)} = -\eta,$$

$$\frac{T'(t)}{c^2 T(t)} = -(\mu + \eta),$$

where we have assumed that μ and η are positive, to avoid exponentially growing solutions as in the string equation.

Thus, we have the equations

$$X''(x) + \mu X(x) = 0,$$
$$Y''(y) + \eta Y(y) = 0,$$
$$T'(t) + c^2(\mu + \eta)T(t) = 0.$$

The corresponding basic solutions are therefore

$$u_{\mu,\eta}(t, x, y) = \sin(\sqrt{\mu}x)\sin(\sqrt{\eta}y)(a\exp(-c(\mu + \eta)t)).$$

We have used only sine terms for the $X(x)$ and $Y(y)$ components because of the boundary conditions, similarly to the string equation expansion. Direct substitution into 10.6 will verify that these are indeed solutions for every μ and η.

Now we want to once again choose a rich enough collection of functions, and this is accomplished by letting $\mu = m^2$, and $\eta = n^2$, giving the solutions

$$u_{m,n}(t, x, y) = a_{m,n}\exp(-c^2(m^2 + n^2)t)\sin(mx)\sin(ny).$$

We would like our final solutions to be of the type

$$u(t, x, y) = \sum_{m=0}^{\infty}\sum_{n=0}^{\infty} u_{m,n}(t, x, y)$$

$$= \sum_{m=0}^{\infty}\sum_{n=0}^{\infty} a_{m,n}\exp(-c^2(m^2 + n^2)t)\sin(mx)\sin(ny). \qquad (10.16)$$

The boundary conditions are automatically met by the choice of the sine functions in x and y. The initial conditions imply that

$$f(x, y) = u(0, x, y) = \sum_{m=0}^{\infty}\sum_{n=0}^{\infty} a_{m,n}\sin(mx)\sin(ny).$$

Thus, we need two-dimensional sine expansions of $f(x, y)$ to find the coefficients $a_{m,n}$.

These can be computed exactly as they were in the case of the 2-D wave equation. The exercises and projects from the wave equation can be extended to the heat equation. The differences are enlightening.

10.3.4   Problems, Exercises, and Projects:

1. Create a small program which will display the evolution of the basic
   waveforms with time with the basic 1-D heat equation, using MATLAB
   or a similar program. The wave should smoothly evolve in time, giving
   one a feeling for its true time evolution.

2. Similarly to Problem 1, let the initial function $f(x, y)$ be a simple
   linear combination of several eigenfunctions, and look at the evolution
   of the function. You may have to adjust the limits of the graph.

3. [**Project**] Write a program that will accept an initial sampled vector
   approximating $f(x)$. Allow the user to input the time steps, and t-x
   resolution. The program should produce a time-evolving graph of the
   wave, as described above.

4. [**Project Heat:2-D**] Write an extended version of the program in
   Exercise 3 which will allow the user to input a discrete, sampled $f(x, y)$,
   in addition to the boundary conditions. The program should produce
   a time-evolving graph of the temperative, which may be more complicated than in 3.

Bibliography

1. Benedetto, J.J.: Harmonic Analysis and Applications. CRC Press (1996)
2. Boyce, W.E., DiPrima, R.C.: Elementary Differential Equations with Boundary Value Problems. Wiley
3. Bracewell, R.: The Fourier Transform and its Applications (1999)
4. Carleson, L.: On convergence and growth of partial sums of Fourier series. Acta Math. **116**, 135–157 (1966)
5. Cheney, E.W.: Approximation Theory. Chelsea Publishing Company, New York
6. Cooley, J.W., Tukey, J.W.: An algorithm for the machine calculation of complex Fourier series. Math. Comput. **19**, 297301 (1965)
7. Feller, W.: An Introduction to Probability Theory and Its Applications, vol. 2. Wiley, New York (1966)
8. Folland, G.B.: Introduction to Partial Differential Equations. Dover Press
9. Gibbs, J.W.: Fourier's series. Nature **59**, 200 (1898); 606 (1899)
10. Herstein, I.N.: Topics in Algebra, 2nd edn. Wiley, New York (1975)
11. http://nobelprize.org/educational/physics/imaginglife
12. http://www.melissamemorial.org
13. Olson, T., DeStefano, J.: Wavelet localization of the radon transform. IEEE Signal Process. (1994)
14. Olson, T., Jaffe, J.: On the effects of squashing in limited angle tomography. IEEE Trans. Med. Imaging **9**(1) (1990)
15. Olson, T., Zalik, R.: On the non-existence of a Reisz basis of translates. Lectures in Pure and Applied Math, vol. 138, pp. 401–408
16. Olson, T.: Optimal time-frequency projections for localized tomography. Ann. Biomed, Eng. (1995)
17. Rudin, W.: Real and Complex Analysis. McGraw-Hill (1974)
18. Skolnik, M.I.: Introduction to Radar Systems, 3rd edn. Tata McGraw-Hill
19. Stanton, A.: Wilhelm Conrad Röntgen on a new kind of rays: translation of a paper read before the Würzburg Physical and Medical Society: 23 January 1896. Nature **253**, 274–276 (1895)
20. Tolstov, G.P.: Fourier Series. Dover, London (1962)
21. Zachmanoglou, E.C., Thoe, D.W.: Introduction to Partial Differential Equations with Applications. Dover Press

© Springer Science+Business Media, LLC 2017
T. Olson, *Applied Fourier Analysis*,
https://doi.org/10.1007/978-1-4939-7393-4

Index

© Springer Science+Business Media, LLC 2017
T. Olson, *Applied Fourier Analysis*,
https://doi.org/10.1007/978-1-4939-7393-4

Printed in the United States
By Bookmasters